Managing the Environment

In the end, we will conserve only what we love,
We will love only what we understand,
We will understand only what we are taught.

Baba Dioum

Managing the Environment
Business opportunity and responsibility

John R. Beaumont
Lene M. Pedersen
Brian D. Whitaker

Butterworth-Heinemann Ltd
Linacre House, Jordan Hill, Oxford OX2 8DP

Ⓡ A member of the Reed Elsevier plc group

OXFORD   LONDON   BOSTON
MUNICH   NEW DELHI   SINGAPORE   SYDNEY
TOKYO   TORONTO   WELLINGTON

First published 1993
Reprinted 1994

British Library Cataloguing in Publication Data
Beaumont, John
   Managing the Environment: Business
   opportunity and responsibility
   I. Title
   658.4

ISBN 0 7506 1574 5

Composition by Genesis Typesetting, Laser Quay, Rochester, Kent
Printed and bound in Great Britain by Clays Ltd, St Ives plc

Contents

Preface

This is a book for managers, policy makers, business and management students, as well as for environmental scientists and engineers. There is an increasing requirement for people to have an understanding of the critical issues about managing our environment. One of the main reasons for writing this book is to attempt to raise the significance of the environmental crises as a strategic management issue.

Our Earth is a complex, finely balanced phenomenon where we all live and work (along with the other animals and plants of the planet). The environment's resources, both renewable and non-renewable resources, are being degraded and depleted by many organizations, which in turn also create pollution and waste. Environmental problems are major and urgent issues facing the Earth and its people.

While this desperate state is not new, the scale of this worsening situation is new and global. It is increasingly recognized as being scandalous. It is connected directly with the industrialization of the last two centuries, especially the post-war industrial reconstruction, associated with an increased material consumption by a rapidly growing population.

For managers and their organizations, doing nothing is neither sensible nor credible. Business may be a major cause of the problems, but it is also a major source of solutions. The current picture in 'developed' countries can be characterized as:

- while the majority of organizations are aware that the environment is of importance;
- a much smaller number of organizations are doing something (some because of real concern and some for public relations, cosmetic purposes);
- an even smaller number of organizations are acting at the strategic business level.

In 'developing' countries, more pressing problems of poverty and illiteracy mean that the environment remains a relatively neglected topic, albeit one linked directly to their development.

Managing our environment should be good business practice, and must be here to stay. The priorities for business are to comprehend how their activities affect the environment and what can be done to reduce or remove the detrimental effects on our planet.

Private enterprise and public action can, and should, work together. The policy debate will continue about the balance needed between market forces and stricter regulations. Business needs to adopt a 'policy and prevention' approach, rather than a 'reaction and repair' approach. Indeed, the objective must be enhancement, rather than mere compliance. There is no suggestion that businesses' focus on growth and profit should be stopped; progress and wealth creation are needed to enable further investments for our environment. Our belief is that businesses must continue to produce profits, because they can provide the only long-term basis for investment towards a sustainable world without poverty. However, we stress the distinction between 'development' and 'growth'. The latter can be both good and bad. Our focus should be on development, exploiting potential, probably by doing things differently rather than necessarily less. The developing issue of corporate governance with organizations having various stakeholders will be of paramount significance for enhanced management of our environment.

This book attempts to fill a gap in a growing literature by providing a business and management perspective with an orientation towards education *for* our environment, rather than education *about* our environment. Traditionally, books have had their origins firmly from roots in the environmental sciences or economics, without extending to the breadth of general management's disciplinary and functional domains. This book is the outcome of a number of years' research and reflection:

- our recent (undergraduate, MBA, postgraduate and post-experience) teaching and learning experiences;
- our consultancy and management backgrounds in the UK, US and Europe;
- our involvement in developing research agendas for the global environment and the funding of associated projects;
- our frustrations with the existing management literature in the field;
- our belief that:
  - the environmental problems are very serious, and effective management, planning and control, of our finite and interdependent planet must commence now;
  - an environmental strategy must be an integral and aligned component of an organization's corporate business strategy;

–   environmental issues will be one of the most important aspects of management in the 1990s and the next millennium.

As a consequence, this book is aimed at indicating:

● how to understand the interlocking environmental crises that are occurring;
● how to comprehend the strategic significance of the environment for businesses;
● how to start questioning conventional wisdom, particularly in the different functional areas of management;
● how to consider an organization's responsibilities for the environment;
● how to evaluate the business benefits to an organization of managing our environment;
● how to place the environmental problems in their broader socio-economic and political context.

One real difficulty with writing such a book is knowing when to finish. The nature of the subject means that there is a constant flow of additional material from enhanced scientific understanding to new policies and modified behaviour. A halt was called, as demonstrated by this book's publication!

As with all books providing a general business and management perspective, it does not offer a detailed discussion of either environmental science or tools and procedures. The necessary 'management', rather than 'technical', outlook is offered. The coverage is broad, and, inevitably, there are gaps. Significant issues are raised and discussed to cause readers, both managers and students, to think and to want to know more! The orientation is towards managers, rather than their organizations per se, in the belief that it is an organization's people that ultimately determine whether opportunities and responsibilities are taken. We believe more environmentally aware managers can only be helpful to societies. Indeed, a recurrent theme is for managers to be proactive in their actions for the environment.

To a large extent this book does not provide any really new material, except an up-to-date and clearer business and management perspective. This lack of basic originality should not be criticized. The recurrent recognition of similar themes and issues means that we are beginning to acknowledge that the real environmental problems will not disappear. Moreover, certain solutions seem to be attainable. At a general level, there is a need for transformations to new ways of working and living. The fundamental problem, however, is not the appreciation of the need for change, but the requirements for agents of change and its subsequent management.

The authors, as a team, are responsible for the contents of this book, and our joint workings have been important learning experiences for ourselves. It is the outcome of the last three years' work, and we have been helped and influenced by many people's ideas. It is impossible to acknowledge everybody individually, but our gratitude to everybody is not diminished. The assistance and encouragement of Jacquie Shanahan and Alison Boyd at Butterworth-Heinemann are much appreciated. Ewan Sutherland deserves a special mention for his careful and questioning reading of an early draft. Our debt to other people's ideas, which we have summarized and quoted, should be obvious. Any errors or misinterpretations, as usual, are the responsibilities of the authors.

John R. Beaumont
Lene M. Pedersen
Brian D. Whitaker

Case studies

1 Can we manage?

Most of today's decision makers will be dead before the planet suffers the full consequences of acid rain, global warming, ozone depletion, widespread desertification and species loss. Most of today's young voters will be alive.

(World Commission on Environment and Development, Brundtland, 1987)

Introduction

Our physical environment, the planet's air, land and water, is a complex, finely balanced phenomenon where we all live and work along with the other animals and plants of the Earth. The environment's resources, both renewable and non-renewable, are being degraded and depleted by many organizations in both the private and public sectors. Moreover, many activities create waste and pollute our environment.

Britain's 1990 White Paper, *This Common Inheritance*, uses a general definition of the environment, quoting from the nineteenth century economist and philosopher John Stuart Mill.

> Is there not the Earth itself, its forests and waters, above and below the surface? These are the inheritance of the human race ... What rights, and under what conditions, a person shall be allowed to exercise over any portion of this common inheritance cannot be left undecided. No function of government is less optional than the regulation of these things, or more completely involved in the idea of a civilised society.

Environmental problems are a major and urgent issue facing the Earth and its people. This desperate situation is not new. Indeed, many ancient civilizations destroyed the environment which gave them wealth. In spite of recent technological advances, the scale of this worsening situation is new and global, and recognized increasingly as being scandalous. It is connected directly with the industrialization of the last two centuries especially the post-war industrial reconstruction, and associated with an increased material consumption by a rapidly growing population. This accelerating consumption is also often unnecessarily inefficient. Our Earth cannot cope.

The management challenge is that 'business-as-usual' is not sustainable in the long term; the fundamental issue is how businesses can be managed to have an improved environmental, as well as corporate, performance. Our focus is on business opportunity and responsibility, although it is necessary to place the discussion in the broader socio-economic and political context at both national and global scales.

The evidence of the environmental crises is clear! While they are described passionately from different ideological stances, they can also be highlighted empirically. This situation has meant the general public's (that is, consumers') environmental awareness and consciousness have increased greatly.

The interlocking crises are neither isolated nor temporary phenomena, occurring on both global and local scales (see Figure 1.1 for a diagrammatic summary of Japan's Ministry of International Trade and Industry (MITI) envisaged policy areas for the 1990s).

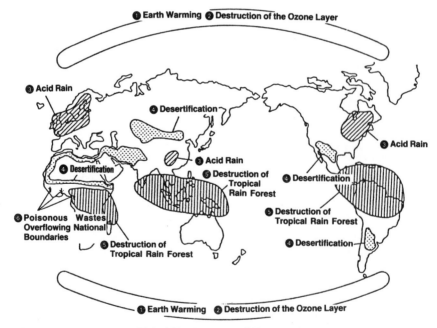

Figure 1.1 Increasing Global Environmental Destruction
(Source : Japan's Ministry of International Trade and Industry (MITI).)

The crises include:

- *Acid rain* The result essentially of industrial pollution, primarily nitrogen oxides, sulphur dioxide and volatile hydrocarbons, that is characterized by its international nature, with adverse impacts on human health, agriculture, fishing and forestry, buildings, ... In the mid-1980s, for example, it was estimated by the OECD to cost approximately one per cent of France's GNP, and ECOTEC estimated the cost of repairing the damage to buildings in the UK to be about £17 billion. More generally, the Brundtland (1987) report argues that acidification in Europe may be so advanced that it is now irreversible.

- *Deforestation* This large-scale activity, resulting in nearly a quarter of the world's rainforests being destroyed in the last thirty years, reduces the Earth's aggregate capacity to absorb carbon dioxide through photosynthesis (and as a process also generates additional carbon dioxide). Trees are essential for the Earth's well-being. They are, however, an important export resource for many 'developing' coun-tries, and this adverse process of deforestation is promoted through market mechanisms, as it was in earlier periods in 'developed' countries.

- *Desertification* New deserts are being created through climatic change and poor management. For instance, the human-induced pressures on the marginal lands of the Sahel from population increase, over grazing, and changes in the nomadic lifestyle have received much publicity with the dislocation of 'environmental refugees'.

- *Global Warming* Scientists believe the increased release of gases, carbon dioxide, chlorofluorocarbons (CFCs), methane and nitrous oxides, from human activities are causing temperature increases. There seems to be certainty about global warming, even if there is uncertainty about its timing, scale and distribution (Figure 1.2 shows the deviations in global mean annual surface air temperatures since 1900, with the reference period being 1951–1970). Estimates indicate that global temperatures will increase by between 1.5 and 4.5°C by the middle of the next century. Consequences of global warming include sea-level rises leading to major demographic displacements and adverse impacts on some industries such as agriculture, fisheries and tourism. Martin Parry (1990), for example, argues that, by about 2030, the mid-latitude 'breadbasket' regions, such as the US Great Plains and the Canadian Prairies, will have 10–30 per cent lower production because of warmer, dryer conditions. Financially, this change can be translated for the US to an annual reduction in income of over $30 billion. A single event, the severe storm of October 1987 in the south of England caused damage of about £1 billion and nineteen deaths.

Centigrade

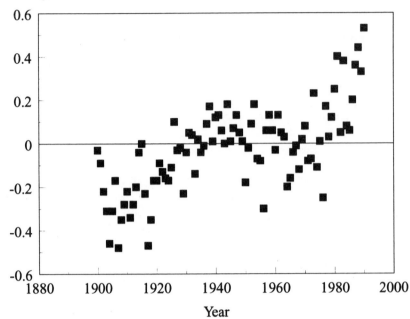

Figure 1.2 Global Mean Annual Temperature Deviations
(Data Source : UNEP, 1991, pages 131–132.)

Remember also the climate tragedies of the Irish potato famine in the
1840s and the US 'dust bowl' of the 1930s.
● *Ozone depletion*   A thin ozone layer protects the Earth from the Sun's
harmful ultraviolet rays. A hole was discovered by Joe Farman of the
British Antarctic Survey in the stratospheric ozone above the Antarctic
in 1985. It has been suggested that chlorofluorocarbons (CFCs) are
destroying the Earth's atmosphere. In 1992, the hole over the Antarctic
had grown so that it is now over inhabited areas, leaving more than
100,000 people exposed to a reduction of up to a half in their protection
against ultraviolet rays. Scientists have also announced that the ozone
layer over Europe has thinned by up to 18 per cent. Evidence of skin
cancer incidences, especially malignant melanoma and eye cataracts,
needs to be monitored.

In addition, it is possible to catalogue a list of much-publicized, large-
scale disasters, including:

● the release of dioxin after a major chemical explosion at Givandan's
plant at Seveso in 1976;

- a leak of poisonous fumes from Union Carbide's plant at Bhopal in 1984;
- the explosion at the Soviet nuclear plant at Chernobyl in 1986;
- the 'NIMBY' (not-in-my-back-yard) syndrome of the 1988 voyage of the polychlorinated biphenyls (PCBs) toxic waste-laden Karin B barge to find suitable treatment capacity;
- the 1989 oil spills into Prince William Sound off Alaska from the Exxon Valdez;
- Saddam Hussein's deliberate oil releases in Kuwait during the Gulf War;
- the 1992 Aegean Sea oil spillage disaster off the north western coast of Spain;
- the 1993 oil spillage from the Braer along the Shetland Islands' southern coastline;

and, unfortunately, many others. The fatal mixtures of bad management, poor design, operating errors and so on have been significant causal factors behind these and other disasters.

Such exceptional disasters should not be forgotten, but it is often the 'everyday', seemingly acceptable, activities at work and in the home that, in total, create the real environmental problems. Wastefulness in our 'paperful' offices, as well as the selection of hardwood furniture, affects the forests, driving to work in part contributes to global warming, and so on and so on. Their cumulative impacts are enormous and growing. The annual Worldwatch Institute's reports provide important summaries of the *State of the World*, and the annual World Resources Institute's world resources reports offer a comprehensive assessment of environmental alterations.

Case: Gaia: Harmony with our living Earth

Named after the Greek goddess of the Earth, in the 1960s, James Lovelock, a former member of the NASA team which considered the possibility of life on Mars and Venus, developed an hypothesis that our planet is a single, living organism. In particular, it is hypothesized that the Earth is a self-regulating system in which all life forms and the environment interact continually and inseparably to ensure life is maintained – holistic geophysiology.

Importantly, the perspective is ecocentric, rather than anthropocentric. The Gaian process of regulation has implications for human activities, specifically our long-term survival without behavioural changes. Briefly, if we continue to delude ourselves that humans have a right to dominate over the rest of the planet (and that science and technology will always assist us), then for self-preservation, Gaia will need to eliminate humanity, replacing it with what James Lovelock terms a 'more amenable species'. That is, ultimately, humans need the planet more than it needs us. This conclusion is identical to the outcome of the economist Nicholas Georgescu-Roegen's (1971) analysis of energy flows and associated economic processes and outcomes. For example, our adverse impacts on climatic change may create new conditions that are not attractive for mankind. In some sense, the removal of biodiversity is death of a kind, because it is completely irreversible.

It is noted that the interest in so-called 'deep ecology', which was introduced by the Norwegian philosopher Arne Naess in the 1970s, also sees a single system with humans having no specific or leading role, although their proposed action is focused on population control rather than broader aspects of development. (See also the views of the Earth First! group.)

With the threat of enormous and uncomfortable structural transformations, James Lovelock (1989, page 177) thinks,

> It follows that, if the world is made unfit by what we do, there is the probability of a change in regime to one that will be better for life, but not necessarily better for us.

US Vice-President Al Gore (1992, page 2) believes,

> ... our willingness to ignore the consequences of our actions has combined with our belief that we are separate from nature to produce a genuine crisis in the way we relate to the world around us.

Earlier, Timothy O'Riordan (1981) classified western environmental ideologies in two ways:

- technocratic;
- ecocentric;

and, for centuries, Aristotle, Francis Bacon, Immanuel Kant and other scientists and philosophers have argued for the recognition of human superiority over other species, with even a right to exert power over others. However, as Lovelock (1979, page 145) stresses,

> The Gaia hypothesis implies that the stable state of our planet includes man as part of, or partner in, a very democratic entity.

Can there be co-evolution of humanity and the remainder of the planet? In considering *The New Realities* in a broad economic, political and social context, the management author Peter Drucker (1989, pages 126–127) emphasizes,

> The final new reality in the world economy is the emergence of the transnational ecology. Concern for the ecology, the endangered habit of the human race, will increasingly have to be built into economic policy. And increasingly concern for the ecology, and policies in respect of it, will transcend national boundaries. The main dangers to human habitat are increasingly global. And so increasingly will be the policies needed to protect and to preserve it. We still talk of 'environmental protection' as if it were protection of something that is outside of, and separate from, man. But what is endangered are the survival needs of the human race.

Three main criticisms of Gaia are usually made:

- the practical difficulties of testing the hypothesis;
- the apparent conflict with Charles Darwin's view of evolution;
- its teleological characteristic with apparently directed changes and a prespecified final state.

It is obviously extremely difficult, if not impossible, to test fully the hypothesis that the Earth is a single living organism, although there are many examples that components do interrelate in a system. The evolution of the environment and the species are assumed to be interdependent, but this does not exclude the particular possibility of Darwinian natural selection. The criticism that the Earth is on a predetermined trajectory is a significant one, because it implies that we should be able to forecast the future and not necessarily do anything about it.

In some senses, however, Gaia has helped to promote environmental issues, especially a holistic perspective on our current unhealthy relationship with the environment, but it does not offer more than traditional systems analysis. Fundamentally, there is no analytical approach to address the problems.

Structure of the discussion

The interrelatedness of environmental issues cannot be exaggerated, and it is their complexities which add significant dimensions to the management challenge, including:

- a need for a multi-functional and interdisciplinary perspective;
- a removal of traditional boundaries between:
  - public and private sectors;
  - countries;
- a recognition of 'natural' and 'man-made' effects;
- a varied interpretation of our environment and its problems, influenced directly by cultural, ecological, economic, ethical, political, social and scientific factors.

It is essential to view the environment as more than an economic issue, associated with resource exploitation and depletion and pollution generation. There are important ecological, political and social dimensions that must be considered by business managers, and economics should be viewed as only one, albeit an important, dimension (see Figure 1.3). The term environmentalism is used to cover the integration of all these dimensions.

There is no instant remedy to these environmental problems; the impulse to act now is laudable and essential, but good intentions do not necessarily guarantee effective action. The environment is a complicated interrelated system, and, while there is an argument that we do need a

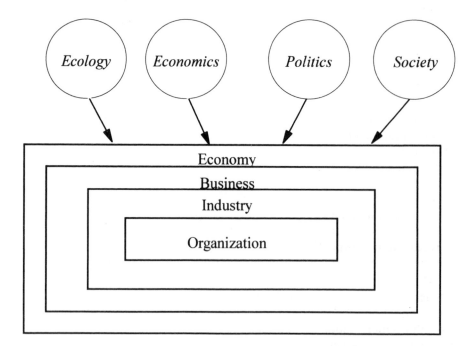

Figure 1.3   Framework for Environmentalism

better comprehension of it, this should not be a case for no action now. Our complacency could be the biggest short-term threat. As US Vice-President Al Gore (1992, page 37) argues,

> . . . a choice to 'do nothing' in response to the mounting evidence is actually a choice to continue and even accelerate the reckless environmental destruction that is creating the catastrophe at hand.

Enhanced awareness, motivation and action by management are necessary, as is an explicit consideration of the political and socio-economic ramifications of our environmental problems. Effective management, planning and control, of our finite and interdependent planet must commence now. Simply stated, finite resources cannot provide the foundation for indefinite growth.

In this book, the argument is developed to demonstrate that business has both a responsibility and an opportunity to assist in managing our environment. Moreover, it is suggested that business actions should become proactive, rather than responsive. Our belief is that businesses, not only industry, must continue to produce profits, creating wealth, because they can provide the only long-term basis for investment towards a sustainable world without poverty. Moreover, we stress the distinction between 'development' and 'growth'. The latter can be both good and bad depending on the environmental pressures it creates. Our focus should be on development, exploiting potential, by doing things differently rather than necessarily less.

As background, in the next section, a brief historical overview is presented to establish the foundation for the contemporary environmental interests across both the political and business communities. The important changes in outlook, from 'no growth' to 'sustainable development' and, hopefully, to 'strategic action', represent a new backcloth for the business and management world to discuss this important topic. In a wider sense, John Young's (1990) examination of *Post Environmentalism* has indicated how existing environmental problems provide opportunities for new political and social structures.

After an examination in the next chapter of the need for the significant issues of the environment to become an integral component of an organization's corporate business strategy, in the following chapters, the discussion is developed around a functional, general management perspective:

- accounting and financial management;
- human resources management;
- information resources management;
- marketing management;
- production and operations management.

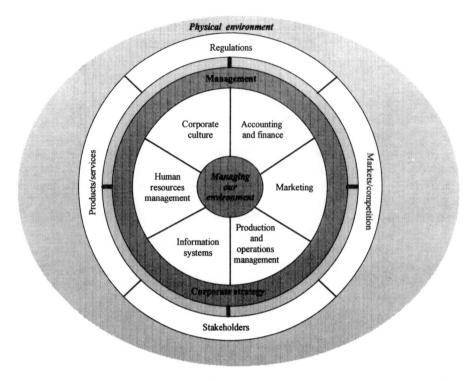

Figure 1.4   Framework for Managing our Environment

Figure 1.4, reflecting the structure of this book, portrays in an idealized form the corporate and interrelated perspective in which an organization should consider the physical environment. By taking a traditional functional perspective to indicate what is required for a sustainable future, adopting sustainable policies and cleaning up the current mess, it is more straight forward to emphasize the need for transformations in management thinking and actions. In a so-called greening of business, many organizations are focusing on specific aspects, often the perceived short-term benefits of cosmetic public relations activities. The environment needs to be recognized as a strategic issue for organizations, with the need for any environmental policy to be cross-functional driven by the senior management. For instance, in Anthony O'Reilly's Chairman's Statement at the beginning of Heinz's 1991 Annual Report, it is highlighted that,

> In May, 1990 an Environmental Audit was conducted across the Company. As a result, an Environmental Team was set up with senior representatives from all functions with the brief to execute a detailed Company-wide environmental strategy for the Company.

Environmental concerns should permeate all aspects of an organization:

- its corporate business strategy;
- its management functions;
- its corporate culture and identity;
- its products and services;

and, additionally,

- its external environment of:
  - competition;
  - regulation;
- its stakeholders.

The complex issues of managing organizational transformation are integral to much of this discussion. More specifically, there are many direct parallels with the current focus on 'total quality management', not the least being that organizations strive for continuing improvement and that management of our environment should be the responsibility of every person in an organization.

At a strategic level, it is becoming more and more important for senior management to be aware of the legal requirements with regard to environmental issues, which increasingly involve international agreements (see, for example, John Slater's (1992) recent guide). Environmental legislation covers a range of issues, including:

- accidents and injuries;
- consumer economic protection;
- dangerous substances and food safety;
- defective products;
- resource and species conservation;
- pollution (air, heat, land, light, noise, radiation, vibration and water);
- safety at work.

It is outside the scope of this book to provide a comprehensive discussion of legislative activity, but it is important to highlight the increasing liabilities that organizations can face. Expert advice is required, for instance in connection with the 1990 UK Environmental Protection Act and the 1986 European Community legal base for environmental action in the Single European Act (Article 130 R, S and T of the Treaty). Linked with the discussion of corporate business strategy in chapter two, there is a discussion of the role of legislation in strategy formulation, and, in chapter three, there is a brief consideration of a range of possible policy instruments, including legislation.

In the final chapter, a global perspective is taken, highlighting particularly both the tensions between developed and developing countries and the implications of the continued population growth. Throughout the book, a forward-looking perspective is adopted, arguing that the environment presents both opportunities and responsibilities for businesses; a 'policy and prevention' approach, rather than a 'reaction and repair' approach is proposed. Moreover, the organizational objective must be enhancement, not merely compliance with existing or future legislation.

Historical overview

In this section, an historical overview of perspectives on managing our environment is presented. While the first indications of the development of an environmental movement came from Rachel Carson's (1963) *Silent Spring*, which was an early questioning of a society's values, specifically the ecological effects of herbicides and pesticides, progress has reflected a slow waltz – advancing, retreating and moving sideways. Our discussion is necessarily selective, and it is organized around the general thinking and actions of the 1970s, the 1980s, and the 1990s, which can be characterized as:

- 1970s: doom and gloom;
- 1980s: sustainable development;
- 1990s: strategic action.

Over this period, there has been an evolving global development perspective, with more attention being given to pollution and waste control as well as to the utilization of scarce natural resources.

1970s: Doom and gloom

At the beginning of the 1970s, a first action by the newly established Friends of the Earth was the dumping of fifteen hundred, non-returnable bottles on Cadbury Schweppes' doorstep. The 'doom and gloom', no economic growth literature documented in various ways the post-war deterioration of our physical environment. In 1972, the United Nations conference on Economic Development and the Environment was held in Stockholm, which led to the establishment of the United Nations Environment Programme (UNEP).

In a special issue of The Ecologist which was published in January, 1972 and entitled *A Blueprint for Survival*, it was argued that,

> The principal defect of the industrial way of life with its ethos of expansion is that it is not sustainable.

Moreover, it was believed that,

> Radical change is both necessary and inevitable because the present increases in human numbers and per capita consumption ... are undermining the very foundations of survival.

In fact, their proposed solution plan was based on zero population growth (with a desire really for a reduction to approximately 30 million people to allow Britain to have national self-sufficiency). The Blueprint's (1972, page 2) case against growth was based primarily on ecological, rather than broader environmental, arguments, and led in Britain to the formation of The Ecology Party, which subsequently renamed itself as The Green Party.

> There is every reason to believe that the social ills at present afflicting our society – increasing crime, delinquency, vandalism, alcoholism as well as drug addiction – are closely related and are symptoms of the breakdown of our cultural pattern which in turn is an aspect of the disintegration of our society. These tendencies can only be accentuated by further demographic and economic growth.

Concerns were also expressed by some industrialists, such as the Club of Rome which sponsored the 1972 report, *Limits to Growth*. This publication attracted much attention, both good and bad. The focus of growth was on industrial output and population, with an expectation that limits to growth would occur within a century. Computer model-based scenarios, comprising a series of equations to represent relationships between variables, were presented to simulate alternative futures. While there were clear messages of potential doom, the real challenge was to explore alternative, long-term and sustainable options. Dennis Meadows and his colleagues (1972, page 24) provided three summary conclusions.

- If the present growth trends in world population, industrialisation, pollution, food production, and resource depletion continue unchanged, the limits to growth on this planet will be reached sometime within the next 100 years. The most probable result will be a sudden and uncontrollable decline in both population and industrial capacity.
- It is possible to alter these growth trends and to establish a condition of ecological and economic stability that is sustainable far into the future. The state of global equilibrium could be designed so that the basic material needs of each person on earth are satisfied and each person has an equal opportunity to realise his or her individual human potential.
- If the world's people decide to strive for this second outcome rather than the first, the sooner they begin working to attain it, the greater will be their chance of success.

Twenty years later, in *Beyond the Limits*, Donella Meadows, Dennis Meadows and Jorgen Randers (1992, pages xv-xvi) argue that their 1972 conclusions remain valid as a conditional warning.

- Human use of many essential resources and generations of many kinds of pollutants have already surpassed rates that are physically sustainable. Without significant reductions in material and energy flows, there will be in the coming decades an uncontrollable decline in per capita food output, energy use, and industrial production.
- This decline is not inevitable. To avoid it, two changes are necessary. The first is a comprehensive revision of policies and practices that perpetuate growth in material consumption and in population. The second is a rapid, drastic increase in the efficiency with which materials and energy are used.
- A sustainable society is still technically and economically possible. It would be much more desirable than a society that tries to solve its problems by constant expansion. The transition to a sustainable society requires a careful balance between long-term and short-term goals and an emphasis on sufficiency, equity, and quality of life rather than on quantity of output. It requires more than productivity and more than technology; it also requires maturity, compassion and wisdom.

In their new, more optimistic examination, the authors give more attention to institutional and technological changes, such as international agreements on greenhouse gas emission levels and CFC substitutes. Briefly stated, the argument is that we are currently living beyond our limits and it is unsustainable, but there are inevitable living choices, rather than a predetermined death sentence. The world is not forecast to end with a big bang, but, during the next century, a large proportion of the world's population will continue personal fights for survival. See Figure 1.5, which presents a summary view of Meadows and colleagues' pessimistic scenario for the twenty-first century based on a continuation of present policies.

With enhanced data availability and quality, significant computing power now readily available, and some methodological advances in dynamical systems theory, the renewed use of large-scale, computer-based simulation modelling is not surprising. Careful interpretation of the results remains of paramount importance, because such analyses can at best provide partial insights into complex problems. For policy analysis, while large-scale simulation modelling is no longer as popular as it was twenty years ago, a potentially significant shortcoming is a lack of financial modelling to complement the modelling of flows of people, pollution, resources and so on. Moreover, while looking at the world as a single entity has obvious benefits, the inability to disaggregate by region or state could be a major shortcoming because it does not allow

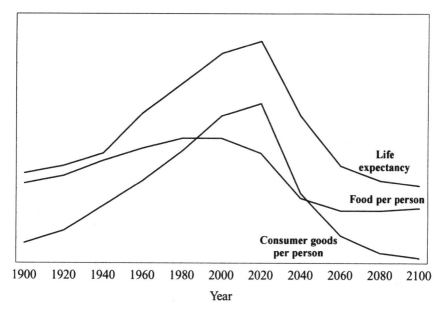

Figure 1.5 Pessimistic Scenario for the Twenty-First Century
(Source: Meadows, Meadows and Randers, 1992.)

some basic questions to be considered. The availability of a desktop computer version of the model, World3, should allow interested people to explore further their own scenarios (although, obviously, within the specific assumptions of the original authors). For instance, to what extent should we assume that the lifestyles of the world's wealthiest countries are desired by all the others? Interestingly, the Dean of MIT's Sloan School, Lester Thurow (1992), has stated,

> If the world's population had the productivity of the Swiss, the consumption habits of the Chinese, the egalitarian instincts of the Swedes, and the social discipline of the Japanese, then the planet could support many times its current population without deprivation for anyone. On the other hand, if the world's population had the productivity of Chad, the consumption habits of the United States, the inegalitarian instincts of India, and the social discipline of Argentina, then the planet could not support anywhere near its current numbers.

While perhaps idealistic and rather abstract, Fritz Schumacher's *Small is Beautiful* received a high circulation for his questioning of the (religious and social) values of the existing economic system. As the full title indicates, *Small is Beautiful : Economics as if People Mattered*, Schumacher's

(1973, page 23) basic orientation suggested the need for decentralization ('small is beautiful') and for the adoption of intermediate technology.

> An attitude to life which seeks fulfilment in the single-minded pursuit of wealth – in short, materialism – does not fit into this world, because it contains within itself no limiting principle, while the environment in which it is placed is strictly limited.

Schumacher (1973, page 11) argued that the fantasy arises because of not differentiating between income and capital.

> Every economist and businessman is familiar with the distinction, and applies it conscientiously and with considerable subtlety to all economic affairs – except where it really matters: namely, the irreplaceable capital which man has not made, but simply found, and without which he can do nothing.

In examining *The Costs of Economic Growth*, Ezra Mishan (1967) argued over twenty-five years ago about 'growthmania', particularly that its maximization has become a fundamental long-term objective of most Governments' economic policies.

At a simple level, unless one has ultimate faith in technological advances, it is not too difficult to assume an insatiable demand for economic growth cannot be satisfied forever by a finite planet; we must have a clear comprehension of the real balance sheet. The problem of managing our environment can, and should, not be separated from the issues of business and economic development; herein, lies both the roots of the problems and the sources of their potential solutions. It is not an over exaggeration to argue that business is the major cause of the problems (at least directly in developed countries and indirectly in developing countries), and business is likely to be the main solution. Wealth creation is essential to enable future investments for our environment.

The perceived 'no growth' sermon had little real effect, primarily because it was felt to be politically and economically naïve. However, many commentators, in fact, were arguing for an examination of developmental options. Until very recently, apparently rational arguments, founded on scientific evidence, have not been sufficiently persuasive to cause political and business action. Over the last quarter of a century, different governments have shown some interest in the environment, but there are many cautionary tales. The 1982 US Interagency Committee, for example, offered an extensive range of facts and figures about resource depletion in their *Global 2000 Report to the President*, but the publisher eventually marketed it with the cover stating *Commissioned by Carter, disregarded by Reagan*!

In Britain, while a Conservative government could be expected to be against no or limited growth policies, in 1974 with a Labour victory, Anthony Crossland, the Secretary of State for the Environment, proved to have no more sympathy for the environmental arguments of the day. His belief was that economic growth should not be constrained because it would permit social development. No politicians believe they will be (re)-elected on a mandate of no growth (see also John Galbraith's (1958) discussion of *The Affluent Society*). Indeed, the economist Wilfred Beckerman (1974, page 239) argued,

> Since it is generally agreed that the progress now being made to reduce birth-rates cannot prevent the world population from doubling by about the end of the century, there seems to be no alternative but to continue economic growth.

In a similar vein, at the 1972 International Labour Conference, the International Labour Office's Director-General's report was entitled *Technology for freedom*.

> We cannot turn our backs on economic growth and technology and innovation on the ground that they are responsible for the deterioration of the environment. Economic growth is the motive power of development, . . . there is no irreconcilable conflict between growth and innovation and environmental protection. We need increased productivity to provide the wherewithal for improving and protecting the environment. This is particularly true in the developing countries, where many of the most serious environmental problems are linked directly to poverty.
>
> (ILO, 1972, page 9.)

1980s: Sustainable development

In 1980, the first Earth Day was held. At present, all countries are a long way from achieving a sustainable future. Anyway, is today's base a sound foundation from which to progress or do we need to repair further? Aurelio Peccei (1977, page 85), founder of the Club of Rome, argued that sustainability and no growth cannot and should not be equated.

> All those who had helped to shatter the myth of growth . . . were ridiculed and figuratively hanged, drawn and quartered by the loyal defenders of the sacred cow of growth . . . the notion of zero growth is so primitive – as, for that matter, is that of infinite growth – and so imprecise, that it is a conceptual nonsense to talk of it in a living, dynamic society.

1987 was European Year of the Environment, although, at that time, from a Director survey, less than half of the managers in the UK had heard about it. It was also the year in which the United Nations' Brundtland

Report, from the World Commission on Environment and Development (WCED), was published on *Our Common Future*. This report was viewed as a sequel to the largely ignored Brandt Report (1980) on North-South relations. WCED was asked to formulate 'a global agenda for change', specifically:

● to propose long-term environmental strategies for achieving sustainable development by the year 2000 and beyond;
● to recommend ways concern for the environment may be translated into greater co-operation among developing countries and between countries at different stages of economic and social development and lead to the achievement of common and mutually supportive objectives that take account of interrelationships between people, resources, environment, and development;
● to consider ways and means by which the international community can deal more effectively with environmental concerns;
● to help define shared perceptions of long-term environmental issues and the appropriate efforts needed to deal successfully with the problems of protecting and enhancing the environment, a long-term agenda for action during the coming decades, and aspirational goals for the world community.

This report, covering common concerns, common challenges and common endeavours, was important because it signalled a philosophical change from the 'anti-growth' perspective of the 1970's environmentalists to one of 'sustainable development' with broader human welfare concerns for the quality and distribution of expansion.

> It is impossible to separate economic development issues from environment issues; many forms of development erode the environmental resources upon which they must be based, and environmental degradation can undermine economic development. Poverty is a major cause and effect of global environmental problems. It is, therefore, futile to attempt to deal with environmental problems without a broader perspective that encompasses the factors underlying world poverty and international inequality.
>
> (Brundtland, 1987, page 30.)

In connection with discussions of our environment, 'sustainable development' is now a widely used, sometimes abused, term. The concept was introduced originally in the International Union for the Conservation of Nature's (1980) *World Conservation Strategy*. It had a focus on ecological aspects, specifically, rather than on broader environmental issues, and, to a certain extent, continued the anti-growth perspective of the 1970s. Sustainable development was given an international, albeit a western, perspective and a more substantive introduction by the Brundtland Commission (1987, page 43), which defines a sustainable society as one that,

> ... meets the needs of the present without compromising the ability of future generations to meet their own needs. It contains within it two key concepts:
>
> ● the concept of 'needs', in particular the essential needs of the world's poor, to which overriding priority should be given; and
> ● the idea of limitations imposed by the state of technology and social organization on the environment's ability to meet present and future needs.

That is, attention should be given to the quality, as well as the quantity, of development and to the conservation of the Earth's natural assets. It must incorporate the removal of poverty and an enhancement in the standards of living among people of the world. It is a global concept that has no meaning for particular countries or specific industries in isolation. To date, the concept has been defined only in terms of general goals, and there remains a lot to be done to develop policies and programmes of actions. It would be incorrect to view sustainable development as some desired fixed end steady-state; it is more a strategy and set of management processes to ensure the long-term potential of the Earth is realized and is not compromised. Sustainable means:

● *not* using non-renewable resources faster than renewable substitutes can be found for them;
● *not* using renewable resources faster than they are replenished;
● *not* releasing pollutants faster than the planet can process them to be harmless.

In a business sense, sustainability, therefore, means income generation while maintaining the (natural and man-made) capital base. The concept of sustainable development is so all-encompassing that its implementation requires:

● a *political system* that secures effective citizen participation in decision-making;
● an *economic system* that is able to generate surpluses and technical knowledge on a self-reliant and sustained basis;
● a *social system* that provides for solutions for the tensions arising from disharmonious development;
● a *production system* that respects the obligation to preserve the ecological base for development;

- a *technological system* that can search continuously for new solutions;
- an *international system* that fosters sustainable patterns of trade and finance;
- an *administrative system* that is flexible and has the capacity for self-correction.

(Brundtland, 1987, page 65.)

A major reason for the environment being on the political agenda in Britain was Prime Minister Mrs Thatcher's 27 September, 1988 speech to The Royal Society, which unveiled a new, somewhat surprising and apparently deep concern with environmental questions.

For generations, we have assumed that the efforts of mankind would leave the fundamental equilibrium of the world's systems and atmosphere stable.

But it is possible that with all these enormous changes . . . concentrated into such a short period of time, we have unwittingly begun a massive experiment with the system of this planet itself.

Furthermore, Mrs Thatcher stated that environmental change in the future,

. . . is likely to be more fundamental and more widespread than anything we have known hitherto.

In considering the need for action, the Prime Minister argued that,

We must ensure that what we do is founded on good science to establish cause and effect.

A landmark in Britain's environmental policy was confirmed by the publication in 1989 of the so-called Pearce Report, *Blueprint for a Green Economy*. In this report, there is a review and summary of the research and debate on the apparent strain between economic growth and environmental protection. David Pearce and his colleagues argue that such tensions are not necessarily incompatible, and outline the Brundt-land concept of 'sustainable development' and how it can be attained through a mixture of market forces and regulations. Given the Government's interest in market-based solutions, it is important to appreciate that manipulation of market forces is not the same as allowing market forces to rule. Development here is interpreted more broadly than 'income', as 'quality of life'.

It is emphasized that a healthy environment is necessary for long-term economic activity, whether it is growing or not. In the preface, David Pearce and his colleagues (1989, page xiv) state,

Sustainable development is feasible. It requires a shift in the balance of the way economic progress is pursued. Environmental concerns must be properly integrated into economic policy from the highest (macroeconomic) to the most detailed (microeconomic) level. The environment must be seen as a valuable, frequently essential input to human well-being. Sustainable development means a change in consumption patterns towards more environmentally benign products, and a change in investment patterns towards augmenting environmental capital.

The authors outline three necessary requirements for sustainable development:

- a proper valuation of our environment;
- an extension to the time horizon;
- a concern for equity, both intragenerational and intergenerational.

These requirements raise basic questions of environmental economics, not only policy issues related to market forces, welfare considerations, decision-making, control and power, but also practical difficulties of forecasting future needs, wants, technologies, and values. The policies that have been introduced have been of two main forms:

- command-and-control', which is founded on regulation, especially standards;
- 'market-based' instruments, including taxes, deposit-refund systems, and tradeable permits.

In practice, a combination of these types of policies are used, although many commentators argue that the latter are more efficient, because they incur lower compliance costs and consumers have an explicit incentive to act in an environmentally conscious manner. Some critics have argued about the socially regressive nature of some taxes, such as energy taxes; however, while superficially true, the revenue raised can be used to assist the disadvantaged in other ways. (For a more extensive discussion, see chapter three.)

In the business community, in the second half of the 1980s, there was increasing awareness and recognition of the significance of environmental issues. The British Institute of Management, the Confederation of British Industry, the Department of Trade and Industry, the Institute of Directors and other organizations began various environmental ('green') initiatives.

At an international scale, the changing political order at the end of the 1980s is also important. Thankfully, the possibilities of a nuclear Armageddon have been reduced by geopolitics. Global nuclear annihilation requires action by people. Inaction, however, will lead to an irreversible global environmental catastrophe. In some senses, the strategic global nuclear threat has been replaced by both the threats to our civilizations arising from environmental pressures and the threats to our environments arising from pressures of human activities. While, understandable, our past concerns have focused on the security of our land, at either national or supranational levels; we must ensure a lasting security for the world's people by considering both environmental and economic development. Resource deployment for 'sustainable development' should help create more peaceful societies. The World Bank (1991) estimate that the military expenditure by developing countries at the end of the 1980s was approximately the same as their aggregate expenditure on education and health (with about a quarter being imports from developed countries).

1990s: Strategic action

The Government's 1990 White Paper, *This Common Inheritance*, presenting Britain's environmental strategy, was criticized for not containing substantive new proposals, but it did bring together in a co-ordinated and comprehensive manner the existing policies. In Britain, the Environmental Protection Act became law on 1 November 1990. One important dimension of this legislation is the introduction of the system of Integrated Pollution Control (IPC) for most industrial processes. Previously, individual components of the environment, air, land and water, have had their own laws for pollution control, even though pollution in one component usually affects the others. Her Majesty's Inspectorate of Pollution is responsible, with associated enforcement powers, for controlling all releases to air, land and water. For specified industrial processes, such as chemicals, minerals and fuel and power, organizations must obtain the necessary authorizations for their activities.

With the progress of environmental concerns being higher on political agendas by the beginning of the 1990s, the failure of it to be discussed, never mind not being an issue for policy statements, during the 1992 British General Election, could be interpreted discouragingly. In party politics, the environment (as well as many other issues) can be deemed ephemeral, depending on the perceived interests of voters at the time. Mike Robinson (1992) argues that, while party ideology has a direct relationship with a party's greening, significant external factors, such as environmental disasters, pressure groups, public opinion and European Commission directives, and internal factors, particularly the actions of individuals, mean the environment is a more significant issue in the 1990s than in the 1970s. There are signs that the subject has become mainstream, rather than a fringe topic, because people no longer dispute the importance of the issues.

However, in his examination of radioactive waste disposal, Ray Kemp (1992) highlights the need for public trust and involvement in the decision-making process, with a new respect for the traditions and values of local communities, rather than the conventional approach of: decide–announce–defend. In the mid-1980s, in Britain, for example, there was the much-publicized four site saga, after NIREX, the government agency, announced the new radioactive waste disposal sites. The co-ordinated reaction was from local communities, local government and some politicians (noting that all four sites were in Conservative constituencies). The outcome was the rejection of the sites, with NIREX then using a deep repository site near Sellafield.

At an international level, the Brundtland Commission appreciated that economic growth has been the cause of much environmental damage, but also argued that economic development is essential to remove poverty. Following the Brundtland Report, in 1989, the United Nations General Assembly planned for the so-called 'Earth Summit'. As highlighted by the much-publicized June 1992 Earth Summit in Rio de Janeiro, otherwise known as the United Nations Conference on Environment and Development (UNCED), which was the world's biggest ever gathering of government leaders and heads of state, there needs to be a serious global commitment to environmental management and sustainable uses of our planet's resources within the context of general economic development and the removal of poverty across the world.

The 1992 Rio conference was concerned about economic development for both developed and developing countries, against a backcloth that the carrying capacity of the Earth can no longer cope with the expected scale of future human activity. The Earth Summit covered three areas:

- a detailed 'blueprint' for implementing sustainable development (so-called Agenda 21) based on 27 principles termed the Rio Declaration;
- a 'framework convention' on biological diversity;
- a 'framework convention' on climate change.

The proposal for an Earth Charter was dropped because it attempted to do too much too quickly. At Rio, there was no discussion of science. It was not the place for such debate, as the underlying science is accepted, at least implicitly. As indicated above, it was the place for debate about the issues by politicians and policy makers, particularly with regard to the estimated required financial commitments and likely institutional procedures.

Where has the Earth Summit left the planet? Notwithstanding the idealism and rhetoric about the environment, Rio was about economic development, in the context of growing, real concerns about the environment. Not surprisingly, the ultimate focus was on money, and the extent to which the allocation of aid would be dependent on environmental actions (and some potential loss of national sovereignty). That is, future aid from developed countries to developing countries will in part be determined by the latter's compliance with environmental agreements; in fact, payments could be made for actions which may not be in their national interest. A fundamental question remains the extent to which environmental management, and indeed economic development and the removal of poverty, can be stimulated by relatively short-term financial mechanisms.

The importance of worldwide economic development and a shared responsibility to preserve the planet's ecosystem are being stressed. A global partnership is envisaged that should not undermine sovereign rights of nations; however, nations must have the responsibility to ensure their activities do not cause damage to the environment of areas beyond the limits of their national jurisdiction.

While it was not surprising that unanimous agreements and actions were not forthcoming, the Earth Summit was more than mere consciousness-raising, and future aid from 'developed' to developing countries is likely to have our planet's conservation as a basic goal. Negotiations prior to UNCED, however, collapsed over who would finance the developments. UN officials have estimated that developing countries will need $125 billion per annum to begin cleaning up their environment and introducing more environmentally sensitive development programmes (noting that aggregate, global military expenditures have decreased enormously in recent years).

While many actions from Rio are still awaited, what potential implications exist for businesses? Although progress will inevitably be slow, not only because of difficulties in obtaining political agreement, but also because of the current world economic recession, the assumption must be that the Earth will affect businesses. Potential impacts can be noted, but their scale and timing are unknown, as there are no agreed specific targets and deadlines. For instance, the biodiversity treaty could have specific implications for the biotechnology and pharmaceutical industries, because of the need to protect animal and plant life and new legal rights for their commercial exploitation. More generally, as David Lascelles (1992) states, a number of other issues were confirmed, including:

- the polluter-pays principle in allocating costs of cleaning;
- more, clear information to be made available on the environmental implications of products, especially with regard to hazardous materials;

- greater control of the international traffic in hazardous wastes, including discouragement of their relocation to countries which cannot really handle them effectively;
- promotion of efficient waste management and recycling.

Prior to the Rio conference, The Business Council for Sustainable Development presented a report on economic development and the environment to Maurice Strong, the Secretary General of UNCED. The main issues of the report, *Changing Course*, are summarized by The Honourable J. Hugh Faulkner (1992, page ii) as,

> We have too many of the same conclusions as others. We have read the many working papers going into the UNCED process from a wide variety of groups – many who have nothing to do with business or industry – and we are impressed at how much similarity exists in the identification and analysis of the issues. The evidence appears to be clear – all those who have thought about these questions of the environment and development are neither divided on what the issues are, nor on the urgency of undertaking change; where we seem to be blocked is by the demand of change itself.

It would be incorrect to give the impression that business has waited until now to consider and act for the environment. For instance, in 1971, IBM became one of the first companies to establish a corporate Environmental Policy, and, in 1984, the first World Industry Conference on Environmental Management was held. In Germany, concern for business opportunity and responsibility for managing our environment led to the establishment in 1984 of the German Environmental Business Management Association (BAUM).

With important business and management inputs into the preparation for the Earth Summit, the Business Council for Sustainable Development argues for a new partnership in changing course toward our common future. They begin their Declaration by stating,

> Business will play a vital role in the future health of this planet. As business leaders, we are committed to sustainable development, to meeting the needs of the present without compromising the welfare of future generations.
>
> This concept recognises that economic growth and environmental protection are inextricably linked, and that the quality of present and future life rests on meeting basic human needs without destroying the environment on which all life depends.
>
> New forms of co-operation between government, business and society are required to achieve this goal.
>
> (quoted by Schmidheiny, 1992, page xi.)

Michael Heseltine, Secretary of State for the Environment, told the Confederation of British Industry's National Conference in November, 1991 that,

> The price of failing the environmental challenge will be that bankers will not lend, insurers will not insure, institutions will not invest, public authorities will not license, politicians will not support, customers will not purchase, and staff will not stay.

Similarly, in their examination of *Stewards of the Earth*, the Institute of Directors (1992, page 3) stresses the role of industry,

> Although industry has widely been cast as the polluter, it is in fact the creator of wealth to raise living standards. Much harm has been done to the environment through governments becoming involved in production processes, especially where they also performed the regulatory function. Good practice in industry is spreading with more and more firms undertaking life-cycle analysis of their products, checking the behaviour of their suppliers and customers. Research and innovation in these leading companies are providing business opportunities. Standards for good environmental management are being developed to encourage boardroom commitment and full disclosure to stakeholders. Training and employment policy must reflect this commitment.

Simply stated, as argued by Sir David Nickson, CBI President, in the Foreword to *Clean up – it's good business*,

> All firms . . . have a responsibility for the environment . . . which can be and, in a number of cases has been, turned to commercial advantage.

In December, 1991, the Confederation of British Industry launched its Environment Business Forum. It is open to all businesses, and it should help them by:

- establishing a national network of organizations committed to the pursuit of environmental excellence in achieving competitive advantage;
- providing up-to-date information on environmental issues;
- demonstrating to legislators the effectiveness of business-led voluntary actions;
- encouraging partnerships between large and small companies to assist the smaller companies to improve their environmental performance.

Participation in the Forum is dependent on demonstrating achievements in the areas covered by the so-called Agenda for Voluntary Action, which has the following criteria:

- designating a board-level director with responsibility for the environment and the management systems needed to address the key issues;
- publishing a corporate environmental policy statement;
- setting clear targets and objectives for achieving the policy;
- public reporting of progress in meeting these objectives;
- ensuring communication with employees and training where appropriate, on company environmental programmes;
- establishing appropriate 'partnerships' to extend and promote the objectives of the Forum, particularly with smaller companies.

Thus, it is becoming clearer to many managers that they can no longer argue or believe that, 'It's nothing to do with me' or that 'I can't afford it'.

At its 1992 Annual Conference, the CBI supported a motion stating that the additional burdens on businesses from developing environmental legislation in Britain and the EC are entirely justified. However, as indicated by the recent extensive discussions of the concept of 'subsidiarity' by politicians of member countries of the European Community, the need for international agreements and legislation is unlikely to be obtained and implemented easily. For example, when Carlo Ripa di Meana and Malcolm Rifkind were the EC Environmental Commissioner and Britain's Secretary of State for Transport, respectively, a row developed over a large project to extend the M3 motorway that would affect the ancient burial mound at Twyford Down. Paddy Ashdown, the leader of the Liberal Democrats argued,

> It comes to something when the Brussels Commission apparently cares more for our environment than our own British Government.

Simply defined, subsidiarity means that action should only be taken at a Community-level when it is more appropriate than action at a national– or regional– level. Environmental subsidiarity would, therefore, restrict EC intervention to those situations that required community-wide action to be effective. This interpretation should not constrain individual countries introducing more stringent legislation than the Community's requirement, although, in certain situations, international trade issues may become involved (see also the GATT discussion in chapter eight and for example the legal judgement about Danish beer in chapter four). While obviously undesirable, it must also be questioned whether environmental subsidiarity could be used as an excuse by particular countries to argue for adherence to standards below the Community's level.

A *Business Charter for Sustainable Development* with sixteen basic principles was launched in 1991 by the International Chamber of Commerce at the Second World Industry Conference on Environmental Management (see Table 1.1), noting that a similarly comprehensive recognition of the need for concerted action had been accepted by the Keidanren's (Japan Federation of Economic Organizations) Global Environmental Charter and its 1990 release of a set of guidelines for Japan's overseas business operations.

Table 1.1   Business Charter for Sustainable Development

1. *Corporate policy*
To recognize environmental management as among the highest corporate priorities and as a key determinant to sustainable development; to establish policies, programmes and practices for conducting operations in an environmentally sound manner.

2. *Integrated management*
To integrate these policies, programmes and practices fully into each business as an essential element of management in all its functions.

3. *Process of improvement*
To continue to improve corporate polices, programmes and environmental performance, taking into account technical developments, scientific understanding, consumer needs and community expectations, with legal regulations as a starting point; and to apply the same environmental criteria internationally.

4. *Employee education*
To educate, train and motivate employees to conduct their activities in an environmentally responsible manner.

5. *Prior assessment*
To assess environmental impacts before starting a new activity or project and before decommissioning a facility or leaving a site.

6. *Products and services*
To develop and provide products and services that have no undue environmental impact and are safe in their intended use, that are efficient in their consumption of energy and natural resources, and that can be recycled, reused, or disposed of safely.

7. *Customer advice*
To advise, and where relevant, educate customers, distributors and the public in the safe use, transport, storage and disposal of products provided; and to apply similar consideration to provision of services.

8. *Facilities and operations*
To develop, design and operate facilities and conduct activities taking into consideration the efficient use of energy and materials, the sustainable use of renewable resources, the minimization of adverse environmental impact and waste generation, and the safe and responsible disposal of residual wastes.

9. *Research*
To conduct or support research on the environmental impacts of raw materials, products, processes, emissions and wastes associated with the enterprise and on the means of minimizing such adverse impacts.

10. *Precautionary approach*
To modify the manufacture, marketing or use of products or services or the conduct of activities, consistent with scientific and technical understanding, to prevent serious or irreversible environmental degradation.

11. *Contractors and suppliers*
To promote the adoption of these principles by contractors acting on behalf of the enterprise, encouraging and, where appropriate, requiring improvements in the practices to make them consistent with those of the enterprise; and to encourage the wider adoption of these principles by suppliers.

12. *Emergency preparedness*
To develop and maintain, where significant hazards exist, emergency preparedness plans in conjunction with emergency services, relevant authorities and local community, recognizing potential trans-boundary impacts.

13. *Transfer of technology*
To contribute to the transfer of environmentally sound technology and management methods throughout the industrial and public sectors.

14. *Contributing to the common effort*
To contribute to the development of public policy and to business, governmental and intergovernmental programmes and educational initiatives that will enhance environmental awareness and protection.

15. *Openness to concerns*
To foster openness and dialogue with employees and public, anticipating and responding to their concerns about the potential hazards and impacts of operations, products, wastes or services, including those of trans-boundary or global significance.

16. *Compliance and reporting*
To measure environmental performance; to conduct regular environmental audits and assessments of compliance with company requirements, legal requirements and these principles; and periodically to provide appropriate information to the board of directors, shareholders, employees, the authorities and the public.

(Source: International Chamber of Commerce, 1991.)

Notwithstanding the increased discussion about business and the environment, empirical evidence from a Touche Ross (1990) survey, however, indicates that,

> Companies ... are underestimating the potential impact of future environmental legislation. They are aware of those problems which get media attention, but miss other environmental problems which impinge upon them .... Companies are aware of some environmental issues but are failing to convert their concerns into positive actions.

Indeed, Michael Heseltine, then Secretary of State for the Environment, criticized major British companies because only one half of them have an environmental policy (Hunt, 1991). Moreover, the majority of measures that have been introduced can be thought of as defensive measures, such as so-called 'end-of-pipe' solutions to existing industrial pollution generation.

In a 1992 KPMG Peat Marwick annual survey of UK business opinion, approximately two thirds of the companies stated that environmental issues were not a trading issue last year, and a slightly lower proportion think the environment will not be an issue in the forthcoming year. Nearly half of manufacturing companies said they were affected by environmental issues, but the corresponding statistic for service companies was only a quarter. Moreover, smaller companies indicated that they are less likely to consider the environment, being mainly affected by local authority requirements rather than national or international legislation.

Britain's Advisory Committee on Business and the Environment, under the chairmanship of John Collins, Chief Executive of Shell (UK) Limited, established in 1991 an Environmental Management Working Group that completed empirical research on environmental standards and economic performance. The results are be summarized in Table 1.2

From recent research on the environmental experiences of small and medium-sized enterprises in Britain, a large range of different actions, with varying degrees of innovation, have been identified, including:

- energy usage and management;
- transport policies;
- waste reduction/minimization;
- recycling;
- product design/product positioning;
- environmental audit/review;
- social responsibility programmes;
- insistence on the achievement of environmental standards by suppliers; and
- public reporting.

Table 1.2

Response categories	Positive responses (%)
Environmental Programmes (What?):	
emission control	58
recycling	75
environmental reviews/audits	82
training	54
energy management	61
waste management	53
environmental policy	68
site improvement	32
supply chain	19
quality initiatives, including BS5750	63
Benefits (Why?):	
bottom line	61
image	65
better relations	40
competitive advantage	28
working conditions	12
survival	23
International Comparisons:	
unfair advantage – domestic laws (or lack of!)	19
unfair advantage – uneven enforcement	35
Promulgation Channels:	
trade associations	60
specialist groups	18
Chamber of Commerce	14
local/national government	16
Confederation of British Industry	19
education institutions	9
suppliers/customers	18
site visits	7
media	16
corporate channels	14
Environmental Legislation:	
use own lawyers to keep up to date	18
use consultants to keep up to date	14
law is best means of enforcing higher standards	7
COSSH is best example of legislation	–
US is worst example of legislation (such as Superfund)	–

Perhaps not surprisingly, no new, instant solutions have been identified. However, in the discussion of the business benefits, four categories can be highlighted:

1 *Tangible benefits*, where the programmes deliver clearly measurable financial benefits. These programmes are most frequently in the energy, transport and waste areas. In most cases, the investment decisions have been justified using traditional cost-benefit criteria, without any special allowances for environmental considerations.

2 *Intangible benefits*, where it is considered that the programmes have delivered benefits, but they are not quantifiable. Typically, these benefits include:
   – competitive advantage, due to the company being seen to adopt a greener position than its competitors. This is very pronounced where companies are selling to the public;
   – employee attitudes, either in the existing work force, or in the ability to attract recruits;
   – relations with the community.
   These intangible business benefits are realized most often from product, social responsibility and public commitment programmes. Warnings are made frequently about the real dangers of these programmes backfiring, if they are thought to be without substance.

3 *Survival benefits*, where the adoption of appropriate programmes is seen as fundamental to the company's survival. These include:
   – the avoidance of future costs, which could be litigation or expenditure on long life capital projects;
   – the necessity to conform to the requirements of major customers insisting on higher environmental standards;
   – conformance with legislation.

4 *Catalytic benefits*, where an environmental review or audit has identified potential cost savings.

From their research, the Advisory Committee on Business and the Environment believes there is a business case for the environment. There are also clear messages for those companies which have not yet begun to implement environmental programmes:

● carry out an environmental review;
● start recycling now;
● start managing energy consumption now;
● define a company environmental policy;
● reduce emissions;
● train your people;
● listen to your Trade Association.

These activities will improve a company's financial performance and its image.

Case: Business opportunity and responsibility: British Gas plc

British Gas, a recently privatized company, is a major energy company, whose products and operations have direct implications for our environment.

British Gas is a founder signatory of the Business Charter for Sustainable Development, which was launched in April, 1991 by the International Chamber of Commerce at the World Conference on Environmental Management (and which is endorsed by the United Nations Environment Programme). The Charter's first principle is:

> ... to recognise environmental management as among the highest corporate priorities and as a key determinant to sustainable development; to establish policies, programmes and practices for conducting operations in an environmentally sound manner.

As company policy, British Gas aims to protect the natural environment and take account of the environmental implications of its operations by:

- complying with the spirit as well as the letter of legislation and approved codes of practice, co-operating fully with relevant statutory and non-statutory bodies;
- assessing the likely environmental effects of planned projects and operations, and maintaining throughout its operations, standards of environmental protection which reflect best industry practice in comparable situations (and improving on such standards where reasonable, practical and economic);
- fostering among staff, suppliers, customers, shareholders and communities local to British Gas operations, an understanding of environmental issues in the context of the business, and reporting publicly on their environmental performance.

Towards enhanced environmental performance, three basic steps have been taken by British Gas:

- initiating, with independent advice, an environmental audit of all activities world-wide;
- strengthening environmental management in the company at the highest level, including the appointment of a Managing Director with specific responsibility for the environment;
- publishing their statement of environmental policy.

The British Gas 1991 Environmental Review was viewed as a 'major contribution' by the Chairman and Chief Executive, Robert Evans, to honouring a company policy commitment to report publicly on their environmental performance.

A number of fundamental challenges exist,

> We need to realign our business thinking and practices to embrace the concept of sustainable development – to devise ways of economic development that can meet present needs without compromising the ability of future generations to meet their needs.
>
> We must reappraise our approach to the environment, and integrate it fully into our strategic thinking and management practices. It is our responsibility to ensure that the necessary policies, structures and procedures for environmental management are put into effect.

In terms of new business opportunities, British Gas's argument is founded on the fact that natural gas is the cleanest fossil fuel (and, therefore, where possible, should substitute for coal or oil) and it can be highly energy efficient. For instance, substitution of petrol by gas as a fuel for motor vehicles would reduce carbon dioxide emissions and should decrease global warming. British Gas has established a management team to specify and develop the natural gas vehicle market. As an increasing global player, their company believes that technology transfer is a basic strategic component. Conversion of Ankara, Turkey to natural gas, for example, should help assist the amelioration of air pollution problems, and, having taken equity positions in the former East German gas industry, the introduction of natural gas should help to reduce the region's heavy pollution.

Conservation of finite resources, through improved energy efficiency, is an integral component of the organization's marketing strategy.

> At British Gas, energy efficiency has always been seen as an integral part of our marketing strategy. Energy efficiency helps to win and retain customers by making gas better value for money. It is also a major policy option for combating the threat of global warming.

In January 1991, for the UK, British Gas launched, through a Code of Practice for handling domestic customers, a new commitment to energy efficiency with a strong orientation to customer awareness and education. The main elements of the Code are:

- energy advice desk with a telephone service in all regions;
- energy labelling on fires and heaters giving advice on running costs;

- staff training and displays and publications in showrooms;
- energy advice service for builders, developers, local authorities and housing associations;
- energy efficiency pilot scheme for low income households;
- maintaining R&D effort on energy efficiency.

While the 1991 Environmental Review is seen as an initial stage in British Gas' evolving approach to environmental reporting, it is introduced by the following statement,

. . . we have taken action, initiated programmes and deployed resources to ensure that British Gas has a responsible approach to the environment.

(Source : British Gas, 1991, Environmental Review.)

Can we manage?

Peter James (1992, page 135) concludes his overview of the corporate response to environmental matters by arguing that,

. . . it is not too fanciful to imagine the next century as one of 'environmental capitalism' in which environmental protection and enhancement is not only a major operational issue and a substantial market, but also a central objective – and source of legitimacy – for both business as a whole and individual enterprises. In this case, the companies taking action now to anticipate environmental opportunities and pre-empt environmental threats may be developing competitive advantage on a timescale measured not in months or years, but also in decades or even centuries.

Today's environmental problems are not simply a matter for science. It is now appreciated that the dreams of a science-created cornucopia of human happiness can only be dreams. Moreover, we must have some clear views of the desired ends before we consider the appropriate means. National and international commitment and co-ordination are essential, with the public and private sectors working together with common goals.

Case: Eco-tourism

Tourism is one of the fastest growing industries, and it has major economic and social impacts on a local area, as well as often direct environmental implications. Growth will create benefits. However, without appropriate planning and management, the scale of this activity can cause enormous pressures on both our natural environment and on our historical heritage.

In 1982, a joint declaration by the World Tourism Organization and the United Nations Environment Programme stated,

> The satisfaction of tourism requirements must not be prejudicial to the social and economic interests of the population in the tourist areas, to the environment, or above all to natural resources which are fundamental attractions of tourism.

However, many sites, including World Heritage sites, are suffering from their popularity, and these pressures are likely to continue.

> Very probably, the greatest earning power of natural habitat in the future lies not in its capacity to produce unique goods, but instead in its capacity to produce unique services. International tourism has been one of the world's growth industries for the past two decades; it has increased tenfold during that period. It now represents 5 per cent of all international trade (World Tourism Organization, 1990).
>
> (quoted by Pearce, 1991, page 199.)

Post-Rio, for example, the Brazilian Government and the private sector have mounted an advertising campaign to resurrect the country's ailing tourist trade. The focus is now on the country as a paradise for 'eco-tourists'. *Natureland Brazil* is a campaign financed by American Express Brazil, Embratur, the country's tourist authority, and Varig, the country's largest airline.

At the November, 1992 international Tourism and Environment conference in London, many speakers, including William Davis, Chairman of the English Tourist Board and Jonathon Porritt, President of the World Travel and Tourism Council, argued for more environmentally aware holidays.

Stop Press:

An unprecedented opportunity has been advertised recently to visit the Kola Peninsular under the White Nights of the summer solstice. The Kola Peninsula is one of the most ecologically devasted areas in Russia, including five highly polluting smelters, a nuclear plant, and one of the largest cities north of the Arctic Circle. The total cost of the three week trip is $2995, subject to air fare changes!

Business leadership and businesses in general have a significant and practical role to play in the policies for and the process of managing our environment. No organization can now sensibly ignore its environmental performance, and there are many opportunities and responsibilities that today's managers must be considering. It will not be merely an issue of performance, but it is likely to be increasingly an issue of survival. While organizations are likely to give more attention to the environment in the 1990s than in previous decades, it is unlikely that their policies and actions will be the same. For example, pollution and waste minimization will be more the order of the day, rather than addressing the problems after they are created.

Review questions

1. In what ways have societies' concerns for our environment changed over the last thirty years?
2. Describe the nature of today's environmental crises.
3. Highlight the main characteristics of the authors' proposed multi-functional framework for managing our environment.

Study questions

1. How should the teaching and learning of management be changed in order to prepare managers for more environmentally conscious management?
2. To what extent will the environment become the significant strategic force affecting the performance of tomorrow's businesses?
3. Consider for different industries, why businesses have both responsibilities and opportunities for managing our environment.

Further reading

A number of newspapers have regular features, as well as special reports, on the environment:

- *Financial Times* (Wednesdays),
- *The Guardian* (Fridays),

and some business and management magazines frequently contain relevant articles, including:

- *Business Week,*
- *The Economist,*
- *Fortune.*

In addition, many academic, business and management journals also now publish papers on the environment.

Brundtland, Gro Harlem (1987) *Our Common Future*, Oxford University Press, Oxford.
The important and influential report by the World Commission on Environment and Development.

Buchholz, Rogene A., Marcus, Alfred A. and Post, James E. (1992) *Managing Environmental Issues*, Prentice-Hall, Englewood Cliffs.
Sponsored by the National Wildlife Federation's Corporate Conservation Council, this casebook provides teaching materials that can be used to explore corporate management of the environment. The cases are:

- *Changing perspectives on the environment*:
  - the Amazon rainforest;
  - Delta Environmental and the advance of the Greens in Europe;
  - the big spill: oil and water still don't mix;
  - save the turtles.
- *Public policy, economics, and the environment*:
  - the auto emissions debate: the role of scientific knowledge;
  - the 1990 Clean Air Act and Du Pont;
  - groundwater contamination: a city with problems;
  - Ocean Spray Cranberries, Inc.;
  - the forgotten dumps;
  - the politics of recycling in Rhode Island.
- *Business and the 'New Environmentalism'*:
  - Oakdale: a success story;
  - Marine Shale Processors, Inc.;
  - Polaroid's toxic use and waste reduction program;
  - Dow Chemical: environmental policy and practice;
  - ARCO Solar Inc.;
  - Ashland Oil tank collapse;
  - Du Pont Freon Products Division.

Elkington, John, Knight, Peter with Hailes, Julia. (1991) *The Green Business Guide*, Victor Gollancz Limited, London.
In the words of the authors, this clearly written book with many simple checklists,

... provides practical, down-to-earth advice on how companies and other organisations can achieve competitive environmental quality standards.

This 'how-to' book is meant to assist managers build environmental excellence into their organization's activities, and highlights the importance of all external influences, such as legislation, consumers, . . . as well as internal issues.

The book provides some excellent 'problem solving aids' with an extensive helpline of useful addresses at the end of each chapter. While the assistance of professional consultants can be an important way forward, there must be senior-level commitment from within an organization, to make real progress.

Fifteen very brief case studies are provided:

- McDonald's: can Big Mac trim its waste?;
- ABB Flakt: 'the world's biggest environmental control company';
- Ferruzzi Group: visions of a living chemistry;
- Albright & Wilson: good intentions end up in the dock;
- The Caird Group: colour me green;
- Glaxo: helping fight off the raiders;
- AT&T: think globally, act locally;
- Dow: we are all product stewards;
- Rhone-Poulenc: zero-waste is not an option;
- British Gas: giving old gas sites the works;
- Novo-Nordisk: a new industrial symbiosis;
- The Body Shop International: body and soul;
- ICI: a letter from the chairman;
- Norsk Hydro: owning up;
- Toyota Motor Corporation: a cross between a skunk and a pig.

Koechlin, Dominik and Müller, Kaspar (eds) (1992) *Green Business Opportunities*, Pitman, London.

This edited volume, subtitled *The Profit Potential* is the outcome of an initiative to introduce environmental management into the INSEAD MBA programme. While broad in coverage, unfortunately, it is also uneven in the quality and depth of the arguments of the different authors.

Given the current interest in Japanese management philosophy and practice, Takashi Adachi's chapter is particularly interesting on the ways in which Japanese companies approach environmental issues and how their comprehension of environmental management is changing.

Schmidheiny, Stephan (1992) *Changing Course*, MIT Press, Cambridge, Massachusetts.

The author is Chairman of the Business Council for Sustainable Development, which had an advisory business input into the preparation for the Earth Summit in Rio de Janeiro. There is agreement that business has an important role to play in managing our environment. Importantly, as reflected in the title, how businesses should act is thought to be fundamentally different from the contemporary situation.

A significant proportion of this volume is given to case studies illustrating successful steps toward sustainable development:

- *Managing change in business*:
    - New England Electric: making energy conservation pay;
    - 3M: building on the success of pollution prevention;
    - Du Pont: The CEO as Chief Environmental Officer;
    - Norsk Hydro: environmental auditing;
    - Shell: human resource development;
- *Managing business partnerships*:
    - Nippon Steel/Usiminas: long-term partnership for sustainable development;
    - ABB Zamech: technology co-operation through joint ventures;
    - Eternit: Technology Co-operation for a safer working environment;
    - S C Johnson: catalysing improved supplier performance;
    - The Chemical Industry: introducing responsible care;
    - Leather Development Centre: promoting best practice;
- *Managing stakeholder partnerships*:
    - Northern Telecom, Mexico: technology co-operation to halt ozone depletion;
    - The Wildlife Habitat Enhancement Council: industry in harmony with nature;
    - Electrolux: designing energy-efficient products;
    - Mitsubishi: co-operation for reforestation;
- *Managing financial partnerships*:
    - Nordic Environment Finance Corporation: financing for sustainable
- development in Eastern Europe;
    - FUNDES: promoting small businesses in Latin America;
    - Fundacion Chile: financing technology co-operation;
    - GE Capital: lending and environmental risk;
    - Jupiter Tyndall: investing in the environment;
- *Managing cleaner production*:
    - Dow Chemical: making waste reduction pay;
    - Harihar Polyfibers: promoting productivity to prevent pollution;
    - Holderbank: making cement with less energy;
    - Ciba-Geigy: designing a low-pollution dyestuff;
    - Con Agra/Du Pont: profiting from recycled waste;
- *Managing Cleaner Products*:
    - Smith & Hawken: promoting products of sustainable forestry;
    - Proctor & Gamble: using life-cycle analysis to cut solid waste;
    - Migros: using life-cycle analysis in retail operations;
    - HENKEL: developing substitutes for phosphates in detergents;

- Laing: energy-efficient housing;
- Volkswagen: recycling the car;
- Pick 'n Pay: retailers and sustainable development;
- ENI: developing a replacement for lead in gasoline;
● *Managing sustainable resource use*:
- ABB: Introducing Clean Coal at Vartan;
- Triangle Limited: energy from biomass;
- EID Parry: integrated rural development;
- Aracruz Celulose: sustainable forestry and pulp production;
- ALCOA: sustainable mining in the Jarrah Forest.

In short, an interesting volume, but it does not specify clear priorities for action and how implementation could occur.

Ponting, Clive (1991) *A Green History of the World*, Sinclair-Stevenson, London.
A concern for civilizations underpins this readable, historical account. The focus is on the changing role of humans and on why and how they have transformed their local, regional and global communities. History can highlight past experiences, particularly destructions; the force of the environment has changed over time, but Clive Ponting demonstrates that it is interesting to explore world history through a 'green' perspective.

> In this wider perspective, it is clearly far too soon to judge whether modern industrialised societies . . . are ecologically sustainable. Past human actions have left contemporary societies with an almost insuperably difficult set of problems to solve.

2 Strategic business management

There is a 'green' train coming down the corporate track. Some businesses will waste valuable resources, both human and financial, trying to derail this fast-moving force. A few will recognise its importance and growing influence and use it as a vehicle to reach new markets. The train is environmental quality, and I maintain that corporations who climb on board will be the ones that have the best ride in the 1990s.

(Patrick Noonan, 1992, *The Corporate Board.*)

Introduction

As the discussion develops in this book, there are some fundamental, underlying issues about which managers should reflect. For instance,

To what extent in the future will 'environmental excellence' be related to and required for organizational performance?

Moreover, in a business environment characterized by increasing deregulation and reregulation, to promote competition and customer service within particular industries, it is necessary to understand,

What are the 'carrots' and the 'sticks' for managing our environment?

and, indeed,

What is the role of Government?

In any consideration of the 'carrots' and 'sticks', market forces, social pressures, financial instruments, and regulation, it is insufficient to focus on,

What to do?

Managers must be able to make the business case for managing our environment, and, therefore, be able to answer the basic questions,

Why?

and, perhaps more importantly,

Why not?

Furthermore, managers must know how to manage their organizations to minimize their detrimental impacts on our environment, and be able to monitor their progress against stated objectives and standards.

What tools and techniques are available to manage our environment?
What measurements should be made?

In summary, simply stated,

What are the organizations' responsibilities for managing our environment?

and,

What business opportunities exist for managing our environment?

Unless managing our environment is viewed as a priority integral to all organizations' corporate business strategies, real progress is unlikely.

As background, in the next section, an overview of strategic management is presented. Some of the well-known ideas of business strategy analysis are described, extending and modifying some of the approaches with regard to environmental considerations. Unfortunately, the management texts covering the concepts, techniques and applications of strategy analysis rarely consider environmental issues. In the third section, attention focuses on strategy formulation, with specific reference being given to some of the techniques that illustrate a portfolio approach and competitive forces. An understanding of business processes and potential business process redesign is essential to derive competitive advantage; value chain analysis is introduced in section four. The concluding comments highlight implementation issues, and provide signposts and a framework for action to make the environment the strategic concern that it merits.

Strategic management

The environmental pressures on businesses obviously vary by their industry, and also their scale and scope of activities. A MORI survey, which was published in *The Financial Times* in July, 1991, described the public perception of different industries' 'familiarity' and 'favourability' (noting that the environment is only one of a number possible factors behind favourability). As portrayed in Figure 2.1 which summarizes industries' positions, the nuclear industry has the lowest favourability, while building societies, electricity, gas and food industries are the most familiar and favourable industries.

All organizations must have some strategic management, examining,

Where is the organization going?

and,

Where should the organization be going?

Fundamentally,

What are the organization's business objectives?

As part of the process of strategic management, management formulates an explicit plan of priorities for action with appropriate resource

Mean favourability

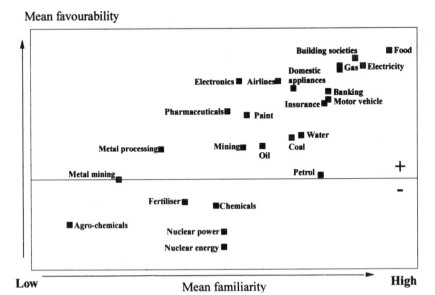

Figure 2.1   Perceptions of Industries' Familiarity and Favourability
(Source: *Financial Times*, July, 1991.)

allocations that will take the organization forward. Particular attention must be given to external trends, including the growing significance of the environment, the changing nature and scope of the industry, technological developments, and the specific competencies of the organization. Situation examinations are completed, such as SWOT (strengths, weaknesses, opportunities and threats) analysis, as well as portfolio analyses to determine an organization's best mix of businesses and of products and services. Strategy formulation is performed usually at different organizational levels, usually:

- corporate;
- business;
- functional.

'Strategy' and 'strategic' have become overused and much-abused terms, in both academia and business alike. Moreover, the topic is often made unnecessarily complicated. In a simple way, Hugh Macdonald captures the underlying and interrelated dimensions of any organization's strategy formulation and implementation through:

- organizational competencies;
- business opportunities;
- corporate governance.

Figure 2.2 portrays a COG diagram with reference to managing our environment.

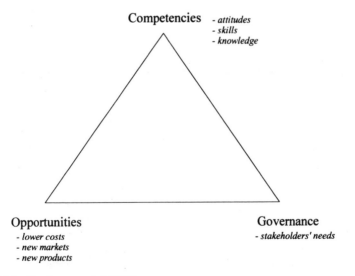

Figure 2.2   Environmental COGs

Management involves the processes of planning and control; strategic management is concerned primarily with the long-term, future direction of the overall organization. Formulation and implementation of strategy cannot be separated, and, indeed, they should be seen as continuous, rather than as one-off, exercises. The essence of strategic management must be action, rather than the process of simply developing the strategy and the document itself. There must be the ability to cope with uncertainty, to devolve responsibility and to retain control. Many managers who devise formal mission statements, specify long-term objectives and prioritize implementation plans, not only waste enormous amounts of time and resources, but also delude themselves into believing that they are managing strategically, even though the strategy is never implemented. In practice, all too often, the real business issues and competitive assessments are not examined sufficiently thoroughly. Such criticisms of the traditional, formal strategic management process could, but should not, be interpreted as abdication of senior management's responsibilities. The essential ingredients are vision, leadership, and, increasingly, a broader comprehension of business's situation in society and its relationship to the physical environment.

The fundamental rationale of the different approaches to strategy analysis is to attempt to explain why some organizations have been successful and why others have failed. A variety of historical analyses has now been completed to examine the observed relationships between organizational performance and different strategies.

Michael Porter (1980) describes three generic strategies:

- *Cost leadership* – at an industry-wide level, an organization can provide an 'identical' product or service at the lowest cost.
- *Differentiation* – at an industry-wide level, an organization can provide a differentiated ('unique') product or service at a premium cost.
- *Focus* – for an identified market segment, an organization can form its product or service provision by either cost leadership or differentiation.

To view strategy analysis as wholly about competition is both unnecessary and unrealistic. Collaboration, inter-organizational linkages, is becoming more and more significant. Moreover, as Michael Porter (1990) demonstrates, much of the traditional theory of comparative advantage is inappropriate to comprehend contemporary international trade. Industries and their organizations are the competitive actors in the global market place. In terms of the Earth's environment, Porter does not believe a sustainable competitive advantage can accrue necessarily from a strategy founded on cheap natural resources, and he argues that firm domestic environmental regulations can provide global competitive advantage for companies in industries affected by regulation. For example, with their strong domestic environmental regulations, German companies supply over two-thirds of the air pollution control equipment sold in the United States.

> German regulation has tended to be demanding and has generally pressured innovation, not impeded it. German environmental standards are also stringent and lead the world in some fields, stimulating innovation in the industries affected.
>
> (Porter, 1990, page 378.)

Case: ESSO's environmental policy

It is the Company policy to conduct its business in a manner that is compatible with the balanced environmental and economic needs of the communities in which it operates. Further, it is the Company's policy to comply with all applicable environmental laws and regulations and apply responsible standards where laws or regulations do not exist. The Company is committed to continuous efforts to improve environmental performance throughout its activities. It will encourage concern and respect for the environment, emphasize every employee's responsibility in environmental performance, and ensure appropriate operating practices and training. The Company will communicate with the public on environmental matters and share its experience with others to facilitate improvements in industry performance.

In furtherance of this policy, it is stated that the Company will:

- work with government and industry groups to foster timely development of appropriate environmental laws and regulations, providing advice on the impact of such laws and regulations on the environment, costs, and supply;
- manage its business with the goal of preventing incidents, and design, operate and maintain facilities to this end;
- respond quickly and effectively to incidents resulting from its operations, co-operating with industry organizations and authorized government agencies;
- conduct and support research to improve understanding of the impact of its business on the environment, to improve methods of environmental protection, and to enhance its capability to make operations and products compatible with the environment;
- undertake appropriate reviews and evaluations of its operations to measure progress and to ensure compliance with this environmental policy.

(Source: ESSO View, 1991, issue 3, page 1.)

Case: Environmental policy in a university

In 1992, an eight-point environmental policy was adopted formally by the University of Essex, England. In a recommendation to the University Council,

> . . . it was noted that many of the suggestions in the policy were already being complied with but it was considered that the whole subject should become part of an approved declaration.

The eight points are:

- minimize any disturbance to the local and global environment and to the quality of life of the local community in which the University operates;
- generally the University seeks to be a good neighbour and a responsible member of society;
- comply fully with all statutory regulations controlling the University and the sites on which it operates;
- maintain the appearance of the University premises to the highest practical standards;
- take positive steps to conserve resources, particularly those which are scarce or non-renewable;
- assess, in advance where possible, the environmental effects of any significant new development and give due consideration to this in formulating its future plans;
- provide the necessary information to enable staff and students to undertake their work, learning or research properly and with minimal effects on man or the environment;
- keep the public informed of major new developments in the operation and expansion of the University.

(Source: Wyvern, University of Essex, 10th June, 1992.)

Strategy formulation and implementation

The process of strategy formulation can vary from an apparently formal and systematic set of procedures for analysis and planning to an ad hoc reaction by the senior management of an organization to perceived problems. As Figure 2.3 indicates, in strategy formulation, managers must consider both internal and external forces. The opportunities and threats in the industry must be considered within the context of the organization's strengths and weaknesses, all of which cannot be divorced

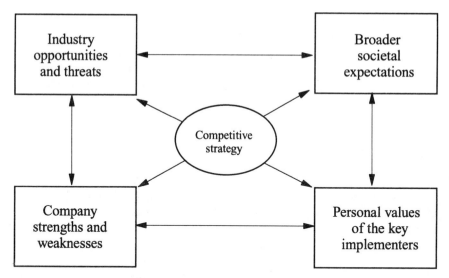

Figure 2.3   Context of Competitive Strategy Formulation
(Source: adapted from Porter, 1980, page xviii.)

from the broader social setting and the leadership and vision of the key
people responsible for implementation.

In terms of business strategy formulation, the regulatory backcloth is
vitally important. The balance between economic instruments and
regulation varies across countries, and this provides the policy context for
business behaviour. As an illustration, it is appropriate to describe the
envisaged future role of environmental legislation in the United
Kingdom, including the requirement for business involvement in its
specification. The political context is described by Michael Howard,
Secretary of State for the Environment, in *This Common Inheritance: The
Second Year Report*, as,

> ... a new presumption in favour of economic instruments rather than
> regulation as we develop our environmental policies.

In the future, the intention will be that,

> ... new regulations should be limited to cases where economic instruments
> to achieve the Government's environmental objectives more effectively are
> either not available or require regulatory underpinning.

In a recent survey undertaken by the Advisory Committee on Business
and the Environment, the general business attitude was summarized
succinctly as,

> If it's good for my business, I'll do it. If not, I'll have to be forced to do it.

Thus, evidence indicates that conformance with legislation is one of business's highest priorities, and that it would be one of the most effective ways, but not the only one, of forcing businesses to do what they perceive to be not 'good' for them. Taking the argument further, activities within a business that could be described as beneficial for the environment are spread across a spectrum:

● obviously *good* for the business;
● less obviously *good* or *not good* for the business;
● obviously *not good* for the business.

In these different circumstances, the motivations for positive business action for managing our environment vary, from business awareness and education being necessary when it is 'good' for business to some form of legislation to force business to act when it is 'not good' for business. In the middle ground, some form of economic instrument may be appropriate, whether it is to bribe or fine a business into relevant action.

As an illustration, to date, as business inputs into the policy debate, the Advisory Committee on Business and the Environment has recommended a number of economic instruments and legislation, including:

● *economic instruments*:
  – energy efficiency;
  – vehicle fuel economy/efficiency;
  – use of public transport;
  – use of diesel fuelled vehicles;
  – company cars;
  – road pricing;
  – cost of landfill;
  – recycling equipment;
  – newsprint de-inking and recycling capacity;
● *legislation*:
  – ban landfilling metal waste;
  – avoid abuse in description of recycled goods.

In addition to the regulatory backcloth, a central concern in business strategy formulation is an examination of the attractiveness of an industry and the appropriate positioning for the organization; such analyses must be dynamic, rather than static. A range of approaches has been devised to complete this evaluation, and two approaches are considered below:

● portfolio analysis;
● competitive forces.

In addition, particular attention is given to a description of a number of different strategic environmental frameworks.

Portfolio analysis

The portfolio approach disaggregates the organization's activities into a set of well-defined Strategic Business Units (SBUs), differentiated by industry attractiveness and competitive position. Individual strategies can be formulated for each business unit, rather than requiring all strategic business units to achieve identical business objectives. The portfolio approach was developed independently by the Boston Consulting Group, McKinsey and Company and the Strategic Planning Institute, although versions, modifications and extensions have been made by many others.

One apparent attraction of the portfolio approach is that it lends itself to simple visual presentation which makes communication relatively straightforward. The basic two-by-two matrix can appear to reduce a seemingly infinite number of potential alternatives into an apparently manageable set of discrete and limited options. The analytical techniques of portfolio analysis have been useful not only for strategy formulation, but also resource allocation and performance specification.

To illustrate the types of techniques in portfolio analysis, it is sufficient to outline the Boston Consultancy Group's:

● experience curve;
● growth-share matrix.

Cost advantage has been stressed in strategy as a primary foundation for competitive advantage in a particular industry. It is also an important issue underlying action or inaction by organizations with respect to the environment. Reasons for cost differentials between competing organizations include scale economies, product design, process innovation, organizational learning and so on; for instance, energy efficiency and waste minimization have provided significant cost reductions for some organizations. The concept of an experience curve is used to explain the changing relative cost structure over time. Derived from empirical evidence, the experience curve maps the inverse relationships between experience (usually defined as the accumulated volume of production) and the average total cost. However, at this aggregate level, no real causality can be implied, although the Boston Consulting Group summarizes the 'Law of Experience' in the chain shown in Figure 2.4.

To provide a visual positioning for each business unit, the Boston Consulting Group developed their well-known two-by-two growth share matrix. The two axes, relative market share and market growth rate, help

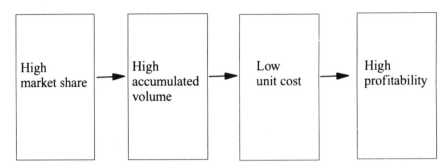

Figure 2.4   Suggested Causality behind the Experience Curve

to differentiate between 'cash generators' and 'cash users', allowing specific strategic actions to be proposed for units in particular boxes. In some senses, market share can be linked to experience (and its relationship with unit costs), and, therefore, other things being equal, it can be related directly to cash generating potential; on the other hand, business growth and environmental management require investments, and, therefore, it can be related directly to cash usage. In the two-by-two matrix, the discrete options have been characterized as (see Figure 2.5):

● *star* – high share and high growth;
● *cash cow* – high share and low growth;
● *question mark* – low share and high growth;
● *dog* – low share and low growth.

This matrix can summarize the expected profit and cash flow for each box, and also indicate an outline strategy: milk the cows; divest the dogs; invest in the stars; and examine the question marks. In their specific consideration of environmentally conscious management, Kaspar Müller and Dominik Koechlin (1992) suggest that the traditional portfolio methods can be supplemented with a control portfolio focusing exclusively on ecological questions. Without developing their argument fully, they highlight the strategic concern that would arise when an organization with a successful product segment identifies it also to be environmentally unsatisfactory.

In contrast, Stefan Schaltegger and Andreas Sturm (1992) argue for the integration of economic and environmental choices when managers evaluate alternatives, whether it be alternative products or production processes. They propose an approach of 'eco-controlling' to assist decision-making, and their so-called 'eco-rational path method' finishes with a portfolio structure to help strategy analysis and development. Their two-by-two matrix (see Figure 2.6) is differentiated by an economic

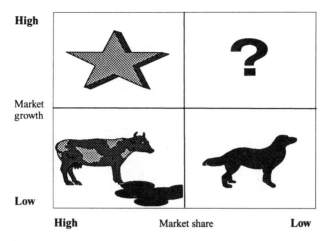

Figure 2.5 Growth-Share Matrix

efficiency criterion, such as the contribution margin, and by an environmental efficiency criterion, such as pollution-flow added; four categories of products are recognized:

- *green cash cow;*
- *black cash cow;*
- *green dog;*
- *black dog.*

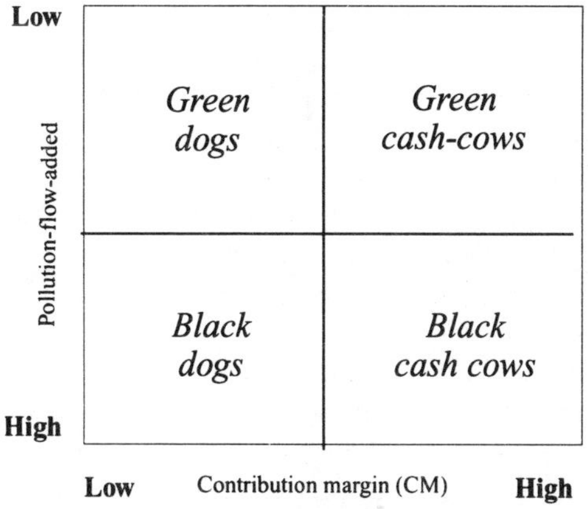

	Low	
Pollution-flow-added	*Green dogs*	*Green cash-cows*
	Black dogs	*Black cash cows*
	High	
	Low　Contribution margin (CM)　High	

Figure 2.6 Portfolio of Eco-rational Path Method
(Source: adapted from Schaltegger and Sturm, 1992, page 237.)

Competitive forces

Michael Porter (1980, 1985 and 1990) has been enormously influential in recent management thinking, especially in the development of competitive business strategies at both corporate and national levels. His consideration of competition in strategic management illustrates aspects of organizational positioning and the factors affecting an industry's overall profitability. Porter's well-known framework of five organization-level competitive forces in an industry can be used to indicate ways in which managing our environment can affect the nature and strength of competition in an industry (see Figure 2.7).

The strength of the forces and the interplay between them affect the potential profitability of an industry and thereby its players. The environment, for instance, has led to new industry entrants, such as the recently nationalized water companies profitably offering a range of waste management services; it has increased the threat of substitute products or services, such as replacements for CFCs and 'green' products generally; it has modified the bargaining power of both buyers and suppliers, as demonstrated by organizations' procurement policies and joint working relationships; and the environment has changed the traditional rivalry in a number of industries, such as the recent promotion of environmentally friendlier motor cars by the manufacturers. While, for convenience of presentation, it is easier to discuss each of the competitive forces in turn, an organization must consider the interrelationships between the forces and act on a number of different fronts. Moreover, the competitive forces of an industry are constantly changing, affected not only by external socioeconomic and political factors, but also by changing internal business and organizational structures.

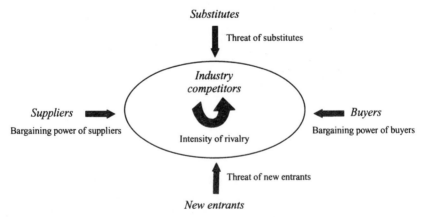

Figure 2.7   Industry Structure Analysis Framework
(Source: adapted from Porter, 1980, page 4.)

Bargaining power of buyers

The bargaining power of both buyers and suppliers affects the possible profit an organization can make in a specific industry. When buyer-power is high, for example when the buyer is large and/or the available products or services are undifferentiated, there are simultaneously downward pressures on prices and upward pressures on both quality and service. Buyer power is also high when there is no real cost to change either to an alternative supplier or to another product. The determinants of buyer power can be divided into:

- *price sensitivity*:
  - price/total purchases;
  - product differences;
  - brand identity;
  - impact on quality/performance;
  - buyer profits;
  - decision makers' incentives.
- *bargaining leverage*:
  - buyer concentration versus firm concentration;
  - buyer volume;
  - buyer switching costs relative to firm switching costs;
  - buyer information;
  - ability to backward integrate;
  - substitute products;
  - pull-through.

Bargaining power of suppliers

The bargaining power of suppliers is great if there are concentrated sources of supply and few substitutes, demonstrated through relatively high prices, susceptible product quality and indifferent service. The determinants of supplier power are:

- differentiation of inputs;
- switching costs of suppliers and firms in the industry;
- presence of substitute inputs;
- supplier concentration;
- importance of volume to supplier;
- cost relative to total purchases in the industry;
- impact of forward integration relative to threat of backward integration by firms in industry.

New environmental pressures, such as an enhanced understanding of the environmental impacts of a product or service throughout its life and more stringent legislation, can change the traditional scale and scope of the bargaining power with buyers and suppliers. Some of the power shifts may be simply that buyers will no longer purchase specific products or be a move to more joint involvement to address environmental problems that cross organizational boundaries and activities.

Threats of new entrants

The problems associated with the threats of new entrants arise because the industry is sufficiently attractive for others to move in. Other things being equal, this causes direct short-term downward pressure on prices from increased availability of products or services. The barriers to entering a market include:

- economies of scale;
- proprietary product differences;
- brand identity;
- switching costs;
- capital requirements;
- access to distribution channels;
- absolute costs advantages:
  - proprietary learning curve;
  - access to necessary inputs;
  - proprietary low-cost product design;
- government policy;
- expected retaliation.

How can increasing concerns for managing our environment be used as an entry barrier to defend a market position or offensively to penetrate the barriers created by others? Some existing organizations, for example, have taken their environmental responsibilities beyond those required by current legislation, partly in anticipation of future regulations and partly because new competitors would have to satisfy more stringent requirements. On the other hand, in some industries, when the existing players have relatively poor reputations for their environmental performance, it may be possible for organizations to enter the industry with newer and more environmentally conscious investments in facilities and processes. Past waste disposal, for example, may result in additional and enormous costs for established organizations.

Threats of substitution

Substitutes (new products or services) can reduce or eliminate the market for existing ones. The determinants of substitution threats are:

- relative price performance;
- switching costs;
- inclination of buyers to use substitutes.

For example, the manual and electric typewriter industry has been almost wiped out by word processing systems. Growing environmental concerns and new regulations requiring stricter standards can influence and perhaps accelerate substitution of products and services by:

- affecting the price-performance relationship of an increasing number of products and services;
- enhancing the functional capabilities of products and services.

The threat of product substitution can also influence entry barriers.

Intensity of rivalry

The intensity of the rivalry between competitors in an industry affects the overall profitability; it is usually greater in mature or declining markets. Following deregulation, for example, the US airline industry has exhibited intense rivalry through price wars, which has resulted in the bankruptcy and take-over of some organizations that were not covering their costs. The relatively new industry of waste management, in comparison, is providing many business opportunities, and some organizations that had to manage their waste effectively as part of their existing operations are now developing profitable services targeted at other organizations.

The rivalry determinants include:

- industry growth;
- fixed costs/value-added;
- intermittent overcapacity;
- product differences;
- brand identity;
- switching costs;
- concentration and balance;
- informational complexity;
- diversity of competitors;
- corporate stakes;
- exit barriers.

Strategic environmental frameworks

To date, perhaps not surprisingly, there has not been the usual spate of different strategic two-by-two matrices to consider management of our environment. It is interesting to note some of the examples. Figure 2.8

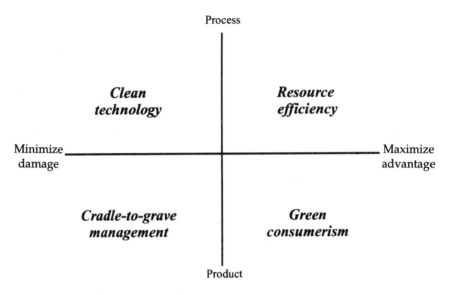

Figure 2.8   Alternative Positions to Achieve Competitive Advantage

suggests alternative competitive positionings related not only to processes and products, but also to optimization of advantages and disadvantages.

Nigel Roome (1992) explores the corporate vulnerability of an organization to environmental matters, which is dependent on the direct and indirect extent to which its activities affect or are affected by the physical environment. Roome proposes two dimensions of corporate environmental vulnerability:

- public perception of environmental impact;
- scientific significance of environmental impact.

Figure 2.9 summarizes the different theoretical positions facing an organization. Nigel Roome (1992, pages 17–18) argues,

> In the case where scientific and public concern about environmental impacts are high, then a company is forced into a reactive position. The company in this context is likely to be increasingly constrained by legislation and public pressure. The response of companies in this reactive position depends on the specific problems they face … the reactive company follows the pace of change dictated by the social, scientific and legislative agenda on the environment.
>
> A similar position is forced on those companies which the public perceive are linked with environmental impacts, even when this is not supported by current scientific evidence. An important distinction between this and the

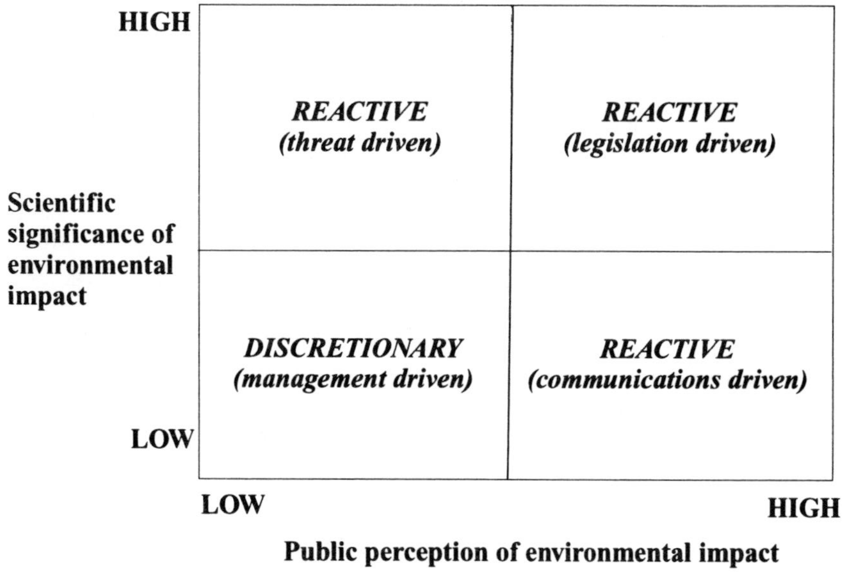

HIGH

REACTIVE *(threat driven)*	*REACTIVE* *(legislation driven)*
DISCRETIONARY *(management driven)*	*REACTIVE* *(communications driven)*

Scientific significance of environmental impact

LOW

LOW   HIGH

Public perception of environmental impact

Figure 2.9 Assessment of Corporate Vulnerability
(Source: Roome, 1992, page 17.)

previous case is that legislative control is less likely to arise in the absence of scientific and public arguments and address the company's presentation of information.

In the sector of the model where scientific evidence about environmental impacts is high and while public opinion remains low, it is possible for a company to adopt a transitional reactive/proactive stance. A company aware of the emerging scientific evidence is reacting to the threat from this information but a proactive response is possible because the company does not yet face pressure from the public and from legislation. The possibility is that scientific concerns, acknowledged within a wider public domain, will shift the company's position to the legislative-driven sector of the model. The actual response may be shallow, involving an assessment of the scientific evidence and communicating the company's position. On the other hand, it may be more fundamental in redirecting the whole company in the light of the impacts it creates.

For comparison in the low perception/low scientific impact sector, their response is discretionary and driven from within the company. Companies in this sector of the model fall into two main categories: those with little apparent environmental impacts and those for which environmental change provides a potential market opportunity. The particular position a company adopts in this sector is chosen by management. The options available include a stance which recognizes incipient threats because scientific discovery and public perception are dynamic. This establishes a need to track environmen-tal, scientific and social change in order to detect emerging areas of environmental vulnerability, market opportunities or to manage the environmental risks and liabilities that are inherent in company acquisitions. With this knowledge it then becomes possible to effect an organizational response as, and when, necessary. A company in this sector can also adopt a complacent position towards environmental concerns, either through conscious choice or by default.

The above interpretation represents a rather pessimistic discussion, with clear signals that a proactive, responsible stance by organizations may not come too easily. Nigel Roome proposes five possible, self-explanatory, strategic options:

- non-compliance;
- compliance;
- compliance plus;
- commercial and environmental excellence;
- leading edge.

Using two, independent dimensions in conventional two-by-two matrices to summarize the strategic issues of managing our environment is obviously rather simplistic. An alternative framework is proposed to help summarize the authors' viewpoint for managing our environment. It is founded on the belief that there is both:

- a business responsibility;
- a business opportunity.

Figure 2.10 idealizes the nature of management's actions for our environment as a spectrum from reaction and repair to policy and prevention.

Business processes

All organizations must understand not only their corporate strategy, but also their business processes for implementation. A business process is a sequence of interdependent tasks and functions that produce outcomes that contribute to the success or failure of an organization; processes describe tasks and not the detail of the work to be done. Simply stated, if all organizations are attempting to provide added value for their customers, it is essential to understand how the organizations create the value. Michael Porter (1985) introduced the value chain framework as a diagnostic tool for understanding competitive advantage and for improving it for an organization.

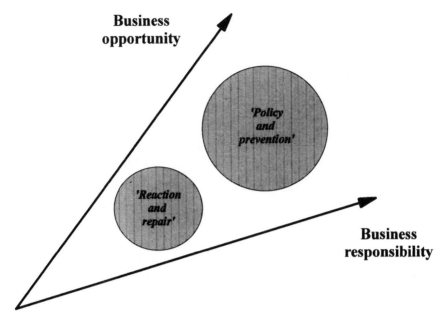

Figure 2.10 Managing our Environment: Business Responsibility and Opportunity

Value chain analysis has two basic dimensions (see Figure 2.11):

- *primary activities*:
  - inbound logistics;
  - operations;
  - outbound logistics;
  - marketing and sales; and
  - service.
- *support activities*:
  - firm infrastructure;
  - human resource management;
  - technology development;
  - procurement.

For managing our environment, Figure 2.12 indicates some of the particular actions that could be undertaken.

By superimposing environmental costs and benefits onto a value chain, it should be feasible to indicate where such costs and benefits are linked to an organization's added value to its customers. Although improvements can accrue within particular parts of the value chain, such as

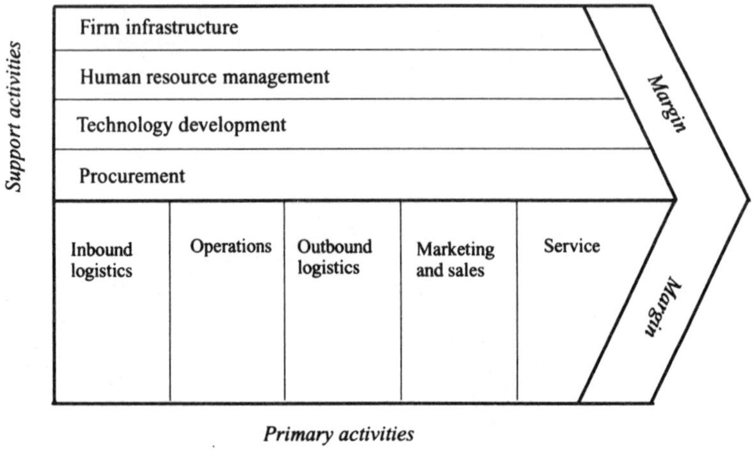

Primary activities

Figure 2.11   Value Chain Analysis
(Source: Porter, 1985, page 37.)

energy efficiency in operations, the analysis framework also emphasizes the linkages across the chain, such as how a change to use recyclable materials can assist marketing of final products. With business process redesign and modified relationships between activities, it will, however, be necessary to attempt to explore such relationships in a dynamic way.

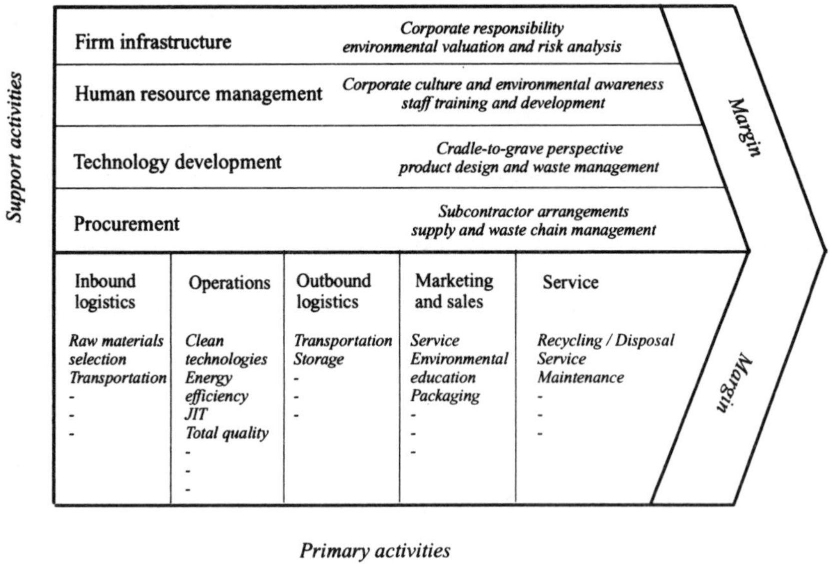

Primary activities

Figure 2.12   Value Chain Analysis for the Environment

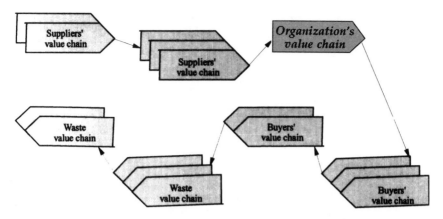

Figure 2.13   Inter-Organizational Collaboration for the Environment

In the car industry, as described in chapter five, the increased attention being given to car design, recycling and disposal is providing management with greater added value for marketing.

Following a collaborative perspective, the value chain can be extended to an inter-organizational perspective (see Figure 2.13). By considering the entire supply chain, or indeed the supply and waste chain, it is possible to consider supplier-buyer relationships as discussed earlier and partnerships for product reconsumption, that is, recycling. For example, IBM, Sainsburys, Marks and Spencer, and B&Q, are some of the increasing number of companies which are putting environmental requirements on their suppliers (as well as assisting them to understand and satisfy new standards).

Peter James (1992) extends Michael Porter's value chain analysis of primary and secondary activities to provide his framework, a so-called 'sustainability octagon', to examine environmental impacts and public acceptability (see Figure 2.14). Value chain analysis recognizes that any product and service must offer added value to a customer, and, in so doing, provide a profit margin to the business organization. James, following Porter's original framework, incorporates five additional components:

● *primary activities:*
  – design;
  – product disposal; and
  – risk management.
● *support activities:*
  – external relations;
  – premises.

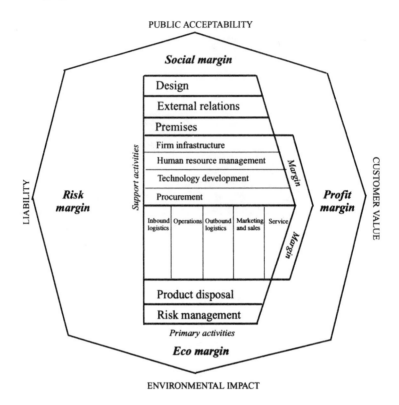

Figure 2.14   Sustainability Octagon
(Source: adapted from James, 1992, page 121.)

In addition, a broader and more responsible business perspective is proposed by including additional, albeit less tangible, margins:

- eco margin;
- risk margin;
- social margin.

Eco margin, for example, is defined as:

> The eco margin is the ratio between an organisation's actual impacts and those impacts which are compatible with sustainable development. Of course, the latter is difficult to calculate, but some proxy measures might be used. In the process area, for example, it might be the ratio between the actual efficiencies and impacts of processes and those which are technically feasible. In the short to medium term, it could be the ratio between present impacts and that required by the most stringent national legislation for each

category of impact. The key point is that, just as a firm cannot operate without a reasonable profit margin, so in the long term it cannot survive without a low, or minimal, eco margin.

(James, 1992, page 120.)

Case: Business opportunity: CFC-free fridges

A company has been saved from going under, by suddenly unveiling the world's first ozone-friendly fridge.

East German firm DKK Schjar-fenstein was facing bankruptcy until three weeks ago after Bosch Siemens pulled out of a take-over bid.

The chief liquidator from the Treuhand was already on his way from Berlin to Saxony to wind up the firm.

Then along came the 'green' fridge. DKK showed their invention to environmental group Greenpeace who promptly stumped up DM100,000 (Ecu 90,000) for a lightning advertising campaign. And within days the company had firm orders for 50,000 of the revolutionary fridges.

Now major international firms, including Electrolux and General Electric, are said to be knocking on DKK's door to talk about a buy-out.

The wonder fridge uses a mixture of propane and butane gases as coolants instead of ozone-destroying CFCs.

DKK had been working on the project for some time and embarrassed Treuhand officials, who have been trying to find a buyer for the loss-making former communist-controlled company for 18 months, say they knew nothing about the invention.

Now they have agreed to absorb DKK's losses for a further three months while talks are held with prospective purchasers who are suddenly keen to get their fingers on a product which could be a world best-seller.

(Source: Andrew Cornwell, 1992, The European, 27th August.)

Concluding comments

Business strategy has only been the subject of systematic analysis for the last three decades. A range of techniques has been developed to assist strategy formulation, but it remains unclear how well these tools will stand the test of time. In a business environment of increasing turbulence and with much greater attention being given to our physical environment, many of the simplistic and rather mechanistic (somewhat deductive) techniques fail to capture the necessarily complex picture of relationships. As complexities are added, the original and attractive clarity of many approaches is undermined. At least, these tools are useful guides, but they do not provide solutions; ultimately, the quality of leadership and management is decisive. In the information age, a greater reliance on detailed empirical analyses, attempting to establish causality, is essential, although, unfortunately, there are not many examples of 'excellent' environmentally performing organizations.

In their discussion of *The Green Capitalists*, John Elkington and Tom Burke (1989) specify ten steps towards 'environmental excellence':

● develop and publish an environmental policy;
● prepare an action programme;
● specify organization, staffing and responsibility;
● allocate adequate resources;
● invest in environmental science and technology;
● educate and train;
● monitor, audit and report;
● monitor the evolution of the green agenda;
● contribute to environmental programmes;
● help build bridges between the various interests.

Real progress still needs to be made in the linkage of strategy formulation and implementation. Although an apparently 'rationalist' approach may be feasible and desirable, learning and intuition should be incorporated. Henry Mintzberg, for instance, has criticized 'deliberate' strategy, arguing that it should be 'crafted'.

Central to this necessary merging of formulation and implementation is clear communication of the strategy with greater involvement and wider ownership across the organization. Moreover, successful strategy implementation requires the business processes, the environmental focus, and the organizational structure to be consistent with, and to reinforce, the strategy.

In this chapter, it has been argued that it is necessary for managing our environment to be considered explicitly as an important and strategic dimension of any organization. While it may be suggested that the environment could be incorporated into the impacts of the various, existing approaches and techniques for strategic business planning, to date, it has not been given the real attention that it merits. Thus, modifications to existing approaches and techniques and the development of new ones are necessary. The environment must be considered systematically by organizations in their corporate business strategy. It is essential for us to be able to look beyond the many clichés in the strategic applications of environmental management. A few heroic management 'gurus' may get by with two-by-two matrices and a lively imagination, but we need tools for everyday managers.

As described in chapter one, the current level of business interest in the environment and associated action varies enormously. While the empirical evidence is not all reassuring, because many organizations are ignoring environmental pressures, it is appropriate to finish this chapter by considering a framework that explores the different degrees of organizational reconfiguration that could be induced by concerns for managing our environment. As Figure 2.15 portrays, six levels of corporate environmental action can be described:

- *ignored* – environmental matters are not considered;
- *localized action* – environmental matters are considered by individual managers and departments, which may include necessary compliance with legislation and measures to improve energy efficiency;
- *corporate action* – environmental matters are recognized to be important and actions result from a broader social responsibility than from a simple preoccupation with short-term profits;
- *business process action* – environmental matters cause an organization to modify its basic business processes;

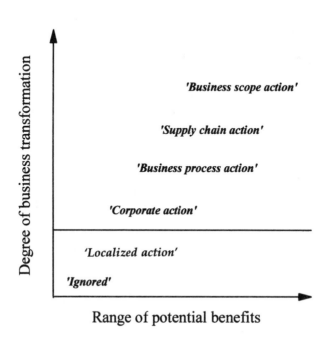

Figure 2.15   Levels Of Corporate Environmental Action

- *supply chain action* – environmental matters cannot be separated easily and co-ordinated action across an industry's supply chain becomes essential;
- *business scope action* – environmental matters allow an organization to expand its activities beyond its traditional business boundaries.

The first two levels can be viewed at best as 'reaction and repair', while the remainder can be described as 'policy and prevention'. The six levels should not be thought of necessarily as a phased approach or automatic developments to increasing management for our environment. However, in the long-term, strategic benefits are only likely to accrue from levels three through six.

Review questions

1. Differentiate between strategy formulation and strategy implementation, highlighting the discussion with environmental considerations.
2. Illustrate the ways in which environmental management can assist an organization's competitive strategy.
3. Comment on the strengths and weaknesses of portfolio analysis, including the suggested extensions to incorporate an environmental dimension.

Study questions

1. Using Michael Porter's framework of five competitive forces, explore the nature of the environmental consultancy industry. Is the framework appropriate, and would it help management's strategy formulation?
2. Examine the assertion that the different approaches really only give a framework for managers to communicate and consider the strategic problems.
3. For a selected organization, discuss the extent, and success, of incorporating environmental issues into its strategy formulation and implementation.

Further reading

Ansoff, H. Igor (1984) *Implementing Strategic Management*, Prentice-Hall, Englewood Cliffs.
   An interesting general text that places the development of strategic management into its historical context.

Grant, Robert M. (1991) *Contemporary Strategy Analysis*, Blackwell, Oxford.

Robert Grant provides a clear and comprehensive MBA-level teaching text which covers the concepts, techniques and applications of business strategy. The integrating theme of the book is 'competitive advantage', including:

- the resource-based theory of strategy (R. Rumelt and J. Barney) with particular emphasis on the role of firm, specific intangible assets (O.E. Williamson, H. Itami) and organizational routines (R. Nelson and S. Winter);
- transaction costs (O.E. Williamson) and their impact upon the scope of the firm (D. Teece);
- value chain analysis (M.E. Porter, McKinsey and Company);
- dynamic approaches to the analysis of competition (J. Schumpeter and J. Williams), including ecological analysis of competitive strategy (B. Henderson, M. Hannan and J. Freeman);
- the creation and sustainability of competitive advantage (P. Ghemawat), including the role of isolating mechanisms and uncertain imitability (R. Rumelt);
- agency problems (M. James);
- reputation effects (D. Camerer and D. Kreps);
- rent appropriation (D. Teece);
- industry recipes (J.C. Spender);
- co-operative strategies and strategic alliances (J. Bower, G. Hamel and C.K. Prahalad);
- the contribution of financial theory to strategy analysis including: value maximization, the role of free cash flow, and option theory (A. Rappaport, M. Jensen and S. Myers);
- transnational approaches to global advantage (C. Bartlett and S. Ghoshal).

Hax, Arnoldo C. and Majluf, Nicholas S. (1984) *Strategic Management*, Prentice-Hall, Englewood Cliffs.

A clear, comprehensive, albeit unexciting, student textbook. The book is divided into four parts:

- the evolution of strategic planning thinking;
- concepts and tools for strategic planning;
- a methodology for the development of a corporate strategic plan;
- the congruence between organizational structure and strategy.

Mintzberg, Henry. (1989) *Mintzberg on Management*, Free Press, New York.

An excellent, clear discussion of 'inside our strange world of organizations'. Many challenges are made about the conventional wisdom of management, based on Henry Mintzberg's experiences of 'bottom-up' management. All MBA students should read Chapter 5, 'Training Managers, not MBAs'!

Peters, Tom J. (1992) *Liberation Management*, Macmillan, London.

The best management book published in 1992! Thought-provoking and practical, but also worrying because how many managers will really be able to cope with the necessary disorganization for the nanosecond nineties?

Porter, Michael E. (1980) *Competitive Strategy*, Free Press, New York.

The most influential strategy book of the 1980s. Porter placed competition into a strategic management context through a presentation of frameworks and techniques. A complementary 1983 volume provides detailed case studies.

Porter, Michael E. (1985) *Competitive Analysis*, Free Press, New York.

Building on the 1980 book, value chain analysis is introduced by Michael Porter as a technique to diagnose and enhance competitive advantage.

Porter, Michael, E. (1990) *The Competitive Advantage of Nations*, Free Press, New York.

In some senses, Michael Porter's most ambitious book (although his Harvard MBAs are forecasting a volume on the competitive advantage of planets in 1995!). This book is beginning to have a major impact on the thinking of management and government policy. As well as clearly developing a fresh socioeconomic approach to a consideration of national economies, there is a large range of empirical investigations that complement and provide insights into the theoretical framework .

As a new framework, Porter argues that the determinants of national competitive advantage can be summarized by the mutually reinforcing systems of a 'diamond' comprising:

- *factor (or supply) conditions*, covering not only traditional factors of production but also important quality drivers such as education;
- *demand conditions*, covering both consumers' consumption patterns and their service expectations;
- *related and supporting industries*, describing the functional clustering of industries;
- *firm strategy, structure and rivalry,* particularly the intensity of domestic competition to ensure organizations are lean and fit.

Scott Morton, Michael (ed) (1991) *The Corporation of the 1990s,* Oxford University Press, Oxford.

A difficult read, providing the summary of the MIT Sloan School of Management's large-scale research programme, 'Management in the Nineties'.

Stacy, Ralph D. (1990) *Dynamic Strategic Management,* Kogan Page, London.

A critical appraisal of the traditional strategic management process, arguing for an explicit consideration of dynamics by striking a balance between opportunism and planning.

Successful strategic management does not mean preparing long-term plans – the future is far too uncertain for that. Instead, it means improving the total control system of the organization so that it is fit and flexible enough to play dynamic business games in highly uncertain environments.

3 Accounting and financial management

In responding to the challenges posed by the environment, which is our natural wealth, all aspects of accounting including financial reporting, auditing, management accounting and taxation will have to change. In doing so, there will be an impact on all members of the (profession) whether in public practice or in commerce and industry and whether working at home or abroad.

(Mike Lickiss, 1991, Accountancy.)

Introduction

There would be no economic problem if there were infinite resources available from the Earth, and they were costed at zero. This situation does not exist, and choices have to be made about current and future patterns of natural resource use. In considering the issue of managing our environment, there is a need, as stressed in chapter one, to go beyond a conventional economics perspective. Economics is supposed to provide systems for allocating scarce resources efficiently. The environment is a set of economic resources, either valuable directly such as clean air and water, or valuable indirectly as inputs to produce other resources such as food and shelter. Notwithstanding the existence of highly regulated and deregulated or reregulated industries, the free market, capitalist system is now the most important system, but 'eco-nomics' fails to incorporate adequately an environmental dimension. Simply stated, the accounting of all environmental costs does not occur, and the necessary investments for the environment are long-term, which appear to be in conflict with today's short-term pressures on corporate performance.

If real environmental costs and benefits could be calculated in full, many evaluations and subsequent decisions and policies would be much more straightforward. However, a number of basic difficulties exists, including:

- there is no market (or, at least, very imperfect markets) for many environmental goods – for example, what price can we put on a beautiful countryside?
- the timescales and uncertainties of environmental impacts, many of which remain unknown, are too long to make conventional discounting of future costs and benefits practical;
- there are welfare and ethical issues of distribution, not only at one point in time, but also between generations.

With such, apparently insurmountable, difficulties over environmental accounting and economics, some people are suggesting an alternative, environmental ethics foundation because many of the issues and problems are inherently different from other business decisions. However, valuation is essential for policy formulation and making decisions about different choices.

In this chapter, aspects of accounting and finance are discussed at a number of levels, particularly national and organizational indicators of economic and environmental performance. Basic issues of valuing our environment, putting full costs on our environmental degradation and depletion, and, consequently, being in a position to reflect the situation with appropriate prices, are considered. The discussion of national indicators, especially the need for alternative measures to the traditional Gross National Product (GNP) and Gross Domestic Product (GDP) statistics, is followed by a similar examination for organizations to incorporate all environmental costs. It is noted that, with our single planet and the fact that environmental impacts are not confined to national boundaries, the discussion could also be extended to an international scale. As a consequence of the national and organizational considerations, attention is given to the envisaged increasing importance of the financial markets for investing for our environment and to issues of social reporting. Some questions are raised about the efficiency and effectiveness of market mechanisms, including the roles and responsibilities of government and the public policy dilemma of reliance on the market and/or on regulation. With the increasing dependence on market-based solutions, rather than command-and-control systems, managers are being given more responsibility for the effective management of our environment.

How do we price the environment?

Some people believe this is impossible (and, as a consequence, stop the discussion at this superficial level). It is probably difficult to view the environment as a 'commodity', but it does possess intrinsic values, which are independent of humans. Clearly, however, individuals and groups of individuals exhibit different preferences for various aspects of our environment. Measurement of preferences is a non-trivial task, but the market-based price mechanisms has proved to be one practical system. A willingness-to-pay questionnaire, for example, is often used to estimate the existence value of say tropical forests, a valuation contingent on their existence rather than their resource usage. An alternative perspective is to consider the opportunity costs from economic activities that do not occur to conserve the scarce resources. Indeed, generally, the real difficulty may be the malfunction or non-existence of an environmental market. It is perhaps not surprising that some activities are not sustainable. For example, some natural resources, such as air, ozone, water, and so on are not traded and are viewed as free, 'public' goods, while other natural resources are grossly under priced. The consequences are excessive current exploitation, with the real costs being borne by future generations.

Appropriate pricing is needed to stimulate efficient resource use and pollution/waste reduction. It must be founded on a comprehensive understanding of all the environmental costs. Unfortunately, such environmental costs, which are termed 'externalities', are often excluded, and, therefore, do not get incorporated into the pricing decisions. These full costs are not only the economic costs of capital, labour and land, and so on, but also the environmental costs of both resource utilization and environmental damage. While some progress has been made to 'internalize' some environmental costs, this must continue even if we have imperfect knowledge and information to make precise estimates. It is likely that progress can be made more easily on identified costs, such as effective waste management, rather than on more complicated issues of valuation, such as for sites of natural beauty or historical heritage.

Any organization that does not understand its costs is unlikely to be successful in the long-term. However, through ignorance and a lack of necessity, managers make decisions related to investments, product pricing and so on that neglect many environmental costs of say natural resource use, pollution and effective waste management. Such costs should probably now include those that must be incurred to repair past damages to the environment. With the fall of the Berlin Wall, for instance, many western organizations are recognizing the opening up of a new, potentially very large market. However, organizations need to be aware that any investments must satisfy European Commission standards and that they could incur liability costs of cleaning up previously inadequate waste disposal.

The concerns in the 1970s about the so-called 'energy crisis', for instance, led to papers on energy pricing, modal split alternatives, relationships between urban layouts and energy consumption and so on. A central actor was the motor car, which has offered enhanced individual and household mobility. *The Machine that Changed the World*, the title of one of the best management texts of the last five years, has also created many well-known problems, including the M25's congestion, Los Angeles's pollution, France's accidents, Tokyo's parking problems, and difficulties associated with rural inaccessibility. Indeed, Paul Ekins and his colleagues (1992, page 34) have asked recently,

What price the car?

The more than 400 million cars in the world impose greater externalities than any other product – at least \$300 billion a year in the U.S. alone. Some specific examples include the following:

- *air pollution:* in OECD countries road vehicles contribute 75% of carbon monoxide, 48% of nitrogen oxides, and 40% of hydrocarbons. World-wide they contribute 17% of carbon dioxide emissions.
- *health:* in the U.S. ... medical costs of pollution from vehicle and industrial fuel combustion are estimated at \$40 billion annually.
- *resource depletion:* in 1986 cars used 200 billion gallons of fuel.
- *congestion:* jams in and around London in 1988 cost \$20 billion.
- *death and injury:* an estimated 250,000 people die in road accidents each year, and over ten times that number are badly injured.
- *blight:* excessive road networks disfigure cities and landscapes, destroy communities, and are the principal areas of danger for children.

National performance indicators

'Economic growth' is defined conventionally as an increase in gross domestic product (GDP), which are statistics derived through so-called national income accounting procedures. National income was the key variable in John Maynard Keynes' (1936) *General Theory*, and such data are needed to manage the economy by his economic concepts and principles. Such statistics were not designed originally to measure a country's progress on broader social and environmental issues, and they are problematic when one wishes to monitor progress towards sustainable development. More specifically, there is a need to define new statistical measurements that reflect the dimensions of 'sustainable development', rather than simply economic growth. (For a more detailed discussion of different indicators specified to assess economics through their environmental and social consequences, see Victor Anderson's (1991) recent proposals for *Alternative Economic Indicators*.) Indeed, in the

1972 *A Blueprint for Survival*, GNP was viewed as an indicator of an economy's environmental pressure.

The traditional measures only include goods and services for which payment has occurred, and changes in their outputs are not necessarily linked to welfare benefits. Unfortunately, our planet is all too often taken for granted as a 'free good' and the finite natural resources are under-priced. Natural capital, such as stocks of animals, energy, plants, and so on, unlike man-made capital, are not depreciated as they are used. Moreover, negative externalities, such as pollution, on the environment that result from economic growth are often forgotten.

> If the air takes away smoke which I created, and I don't have to pay for using the air, then I am getting a benefit from 'environmental wealth', and that benefit will not be reflected in the GNP figures. If the effect of the smoke I create is to reduce the quality of the air, and therefore to reduce the benefits which other people derive from it, that cost – an example of 'environmental depreciation' – will not be reflected in the GNP either.
>
> (Anderson, 1991, page 26.)

To consider progress towards sustainable development, rather than simply economic growth, it may be appropriate to add and subtract different factors to provide an 'adjusted' national product statistic. For instance, Frances Cairncross (1991) argues that it is important that the principle of depreciation is extended to cover the depletion and degradation of natural resources, as well as man-made capital. Following conventional accounting practices, by valuing depreciation, a clearer understanding of the level of necessary replacement investment should be forthcoming, as well as the real costs of business activities. If countries were managed like companies, asset depreciation would be accounted for explicitly. However, with Britain's discovery of North Sea oil, there was no increase in the nation's asset base, and, as it is being exploited, the revenue is included in the nation's income stream without any indication of the associated reduction in (natural) capital.

While 'growth' is easier to measure than 'development', the United Nations Development Programme has attempted to integrate a country's progress through a human development index, linking life expectancy, literacy and gross income. Table 3.1 summarizes their country rankings by the human development index and gross national product.

Frances Cairncross (1991) describes how environmental accounts have been developed by three, distinct approaches:

● physical account;
● sustainable income;
● net domestic product.

Table 3.1 Ranking of Countries by
Human Development Index and
Gross National Product

Country	Human development index	GNP capita
Japan	1	5
Sweden	2	6
Switzerland	3	1
Netherlands	4	14
Canada	5	7
Norway	6	3
Australia	7	17
France	8	12
Denmark	9	8
United Kingdom	10	18
Finland	11	10
Federal Republic of Germany	12	11
New Zealand	13	22
Italy	14	19
Belgium	15	15
Spain	16	26
Ireland	17	24
Austria	18	13
United States	19	2

(Source: UNDP.)

Physical environmental accounting, which is similar to company financial accounts' conceptualization of an opening resource balance that is
modified by stock additions and subtractions through the period of
interest, has been introduced in a number of countries. It is longest
established in Norway, starting in 1974, where there is a differentiation
between:

● *biological* resources;
● *environmental* resources;
● *inflowing* resources (solar radiation, wind, . . .);
● *mineral* resources.

France's so-called Natural Patrimony Accounts are more detailed, with disaggregation of the national picture by region and by organization, in both public and private sectors. Instead of a physical base, the other two approaches have a financial foundation, and, therefore, they can be linked more directly to traditional national income statistics. To give a clear indication of sustainability, it is necessary to depreciate both natural and man-made capital. A 'net domestic product' is calculated by using market values for the exploited natural resources. David Pearce and his colleagues (1990), for instance, talk generally about a 'constant capital rule', recognizing possible trade-offs between natural and man-made capital. However, they also appreciate that some environmental assets are 'critical capital' for which no trade-off would be acceptable. In terms of our long-term environment, a strong argument can be made that confirms non-renewable resources with no direct substitutes are valued as sacrosanct. David Pearce (1993), for example, illustrates the valuing of preferences for unique assets by considering the Grand Canyon.

Victor Anderson (1991) proposes a new statistic, ANP, adjusted national product, which is defined as,

gross national product
minus capital depreciation
plus   money value of unpaid domestic labour
plus   money value of non-money transactions outside the household
minus environmental depreciation.

Conceptually, this modified statistic has attractions, but operationally there are difficulties in providing an unambiguous definition of 'environmental depreciation' and collecting appropriate data. However measured, this cost must go beyond the depletion and degradation of natural resources, and include the expenditures on environmental protection and repair. There is a conceptual attraction of being more consistent than using conventional national statistics, but the operational issues in practice may be significant. At best, however, it gives an aggregate snapshot. It does not capture significant distributional issues and does not reflect the dynamics of sustainability over time. Unless such statistics are derived and published, it is unlikely that governments will have an overall understanding of the environmental impacts of human activities and of their growth policies. Using the definition of a sustainable economy as one that saves more than the depreciation on its man-made and natural capital, David Pearce (1993) and a colleague consider the sustainablity of selected national economies. While recognizing the simplicity of their approach, it is interesting to summarize their conclusions:

● *Sustainable economies:*
  – Brazil;
  – Costa Rica;
  – Czechoslavakia;
  – Finland;
  – Germany (pre-unification);
  – Hungary;
  – Japan;
  – Netherlands;
  – Poland;
  – USA;
  – Zimbabwe;
● *Marginally sustainable economies:*
  – Mexico;
  – Philippines;
● *Unsustainable economies:*
  – Burkina Faso;
  – Ethiopia;
  – Indonesia;
  – Madagascar;
  – Malawi;
  – Mali;
  – Nigeria;
  – Papua New Guinea.

Notwithstanding obvious practical difficulties, some countries have made progress in national resource accounting. Interestingly, in 1973, a new statistic, Net National Welfare (NNW), was introduced in Japan to modify national income statistics for environmental and other factors. David Pearce and his colleagues (1990) have shown that, for the period 1955–1985, while Japanese GNP increased by a factor of 8.3, NNW increased by only 5.8. Germany has recently begun a ten year project to develop another indicator, gross ecological product, which will attempt to incorporate environmental aspects and also be consistent with traditional, economic accounting practices.

A recent UK Department of the Environment (1992) report on *The UK Environment* estimates that the cost of environmental protection is about £14 billion, approximately 2.5 per cent of GDP. About one half of these costs are accounted for by cleaning water, and about one quarter for by reducing different forms of waste pollution. David Pearce (1993) considers provisional, non-official adjustments to one sector of the national accounts, agriculture and forestry. By adding positive environmental effects or benefits to GNP and by deducting negative effects, a significant, upward revaluation of approximately one quarter was estimated. More generally, the Organization for Economic Cooperation and Development (OECD) estimates environmental costs in 'developed' countries to be of the order of three to four per cent of GNP. As the recent growth in GNP has been less than this estimate, it implies that we are borrowing from reserves to give a level of economic growth that does not match the damage.

As mentioned in chapter one, at the Earth Summit in Rio de Janeiro, various concerns were expressed about the planet's declining bio-diversity. Human activities have not only reduced significantly the biological diversity, but also taken an increasing proportion of the Earth and its natural environment. From an ecological perspective and looking at accounts in a consistent unit for considering sustainability, Net Primary Production (NPP) is the aggregate of biological material using the Sun's energy. Although humans are only one amongst millions of species, they account for over 40 per cent of the Earth's products of photosynthesis (and, obviously, they constrain the requirements of other competing animals, and plants). Vitousek and colleagues (1986) estimate that, of this 40 per cent, only one tenth is used directly by humans (see also Table 3.2 and Ehrlich and Ehrlich (1991) for a broader discussion). That is, 36 per cent is taken from other species, either indirectly, such as unused produce returned to the Earth, or as losses, such as asphalted soil. This situation means that maintaining the current diversity is becoming more difficult.

Table 3.2   Human Appropriation of the Products of Photosynthesis

Use	Share of NPP (%)
Direct Use	4
Indirect Use	26
Losses	10

Some form of environmental accounting can provide a framework for data collection to give an aggregate picture of a nation's environment assets and their changes over time, which should give some insights into sustainability. Thus, while GNP statistics indicate countries' economic growth within a particular frame of reference, it is becoming clear that it creates an illusion about countries' development because of a failure to incorporate environmental and social costs.

Case: What are we measuring?

Concerns about traditional economics-based statistics for national income accounting have led to many commentators proposing alternative indicators for our progress in a broader social and environmental context. Importantly: What are these different indicators showing?

Victor Anderson (1991) selected data (1970, 1980, 1985 and the latest available) from fourteen countries, seven 'developed' countries (the so-called Group of Seven) and seven developing countries (the ones estimated to have the largest populations at the millennium). While the use of GNP/GDP statistics indicates general improvement over this period, the use of alternative indicators highlights a more complicated picture.

Specifically, in terms of the environment, Anderson differentiates between 'causes' indicators (population growth, energy efficiency, . . .) and 'effects' indicators (carbon dioxide emissions, tropical deforestation, nuclear reactors, . . .). He concludes that, while the 'causes' indicators are showing a general improvement, the 'effects' indicators show a clear deterioration.

More generally, Victor Anderson (1991, pages 91–93) argues that there seems to be,

. . . three main factors at work, each on a different timescale:

- Social conditions are generally improving, and in the short-term, this is likely to continue.
- In the medium term, environmental deterioration threatens to put these social improvements into reverse. For example, growing desertification threatens current improvements in calorie supply; pollution will threaten current improvements in health.
- In the long term, the outcome depends on whether the current improvements in environmental 'causes' indicators (such as energy intensity and rate of population growth) continue and are on a sufficiently big scale to put the environmental 'effects' indicators into reverse. This would allow the social indicators to resume their past trend of general improvement.

The key differences between this conclusion and the views of the 'anti-growthists' and 'pro-growthists' . . . are as follows:

- The anti-growthists usually want to play down the significance of social improvements. The effect of this is to take away any sense that the world faces genuine dilemmas, and to make the issues appear much simpler than they really are. Just putting current economic and social development basically into reverse (assuming sufficient political support could be won for doing so) would have enormous costs. A more realistic project (not only in terms of political support, but also in terms of what the situation requires) is the attempt to combine as many of the human advantages of 'development' as possible with as few as possible of the environmental and human disadvantages. This will be a complex task, but, hopefully, not an impossible one.

● The pro-growthists' use of GNP and GDP growth rates gives a far too optimistic picture of the world. It plays down both distributional issues (about inequalities in health, wealth and consumption) and environmental issues. Some indicators – such as illiteracy rates and access to safe drinking water – do show a close connection with GNP per head, but others do not, and some environmental indicators show a negative connection, that is, prosperity bought at a high environmental cost. Though the pro-growthists rightly draw attention to benefits from increasing prosperity, such as falls in infant mortality rates, a continuation of economic growth in its present form into the future would be disastrous, because of the environmental costs involved. Statistics for the main indicators used by economists – GNP, balance of payments, inflation, and so on – do nothing to convey the nature and seriousness of the situation we face.

Organizational performance

It is necessary for the traditional measures of corporate economic performance to be extended to include their environmental performance. While internalizing full environmental costs should be beneficial for the environment (although not necessarily for individual organizations and industries), there will need to be a fundamental change in philosophy from a focus on short-term profitability to longer-term measures of return on investments.

One of the reasons for the lack of business investments for the environment is the large, up-front costs that are often involved. However, as with many investments, in practice, evaluations of the costs and benefits of any investment are not straightforward. For investments concerned with managing our environment, the task is admittedly especially difficult, but it is necessary increasingly to undertake it. To a certain extent, the difficulties are related to the failure to fully account for all environmental costs, and the absence of relevant techniques. Moreover, as well as an initial investment appraisal, there needs to be a proper monitoring of any investment. Many people argue that environmental issues are basically different from other management decisions, like the impossible difficulties of valuing human life.

In this section, the basic objective is to raise the neglected issues, rather than provide a tool-kit of financial techniques. Having stressed that the environment should be given a strategic position by organizations, it must also be emphasized that any investment for the environment should be judged on financial grounds. The financial appraisal must explicitly recognize the particular nature and characteristics of an investment for the environment, noting that traditional cost-benefit analysis is generally both inappropriate and inadequate. Organizations must be able to prioritize their investment decisions, environmental and non-environmental ones, because of their need to manage scarce resources, particularly finance and human resources.

For comparative investment analysis, the concept of discounting over time is significant. In discounted cashflow analysis, for example, attention is given to future streams of income in relation to the value of money today. High discount rates are usually used to indicate high risks. In addition, it is important to consider an investment's payback period, the time required to pay back the initial costs of the investment.

With environmental returns likely to be of a long-term nature, traditional discounting concepts are unhelpful and existing interest rates will always be too high. However, following the Brundtland Report's differentiation between intra– and inter–generational requirements for resource utilization, it is suggested that conventional discounting is unsuitable for preserving future generations' welfare.

It is noted that organizations are being encouraged to incorporate systematic decision processes that take the 'best practicable environment option'. For instance, the concept of the Best Available Technology Not Entailing Excessive Cost (BATNEEC), which originated from the European Commission's concern to link the latest technology to pollution standards, is important in bringing environmental considerations into investment appraisals. The 1990 UK Environmental Protection Act requires companies to adopt BATNEEC.

For the environment, the fundamental issue relates to the time horizon that should be adopted. Discounting tends to undervalue longer term decisions, whether new benefits accrue and/or new costs and liabilities arise. As discussed in a number of chapters, the structure of investment appraisal, as well as the short-termism of corporate performance, may be a significant factor behind investments in so-called 'end-of-pipe' solutions which attempt to deal with pollution problems after they have been created, rather than investments in new operational processes to remove the problems before they are created.

In policy terms, for twenty years, there has been agreement to the 'polluter pays principle', which means that a polluter pays the full costs of environmental damage caused by its production of goods and services. In one sense, the principle is mistitled, because, as organizations pass on their costs to consumers through their pricing policy, ultimately, it is the consumers who pay. This situation is sound, because, in the market system, consumer choice is an important driver. This principle, whether consumers pay additionally in the end, is aimed at internalizing some of the environmental costs that an organization generates.

Simply stated, for progress toward sustainable development, environmental criteria and costs will need to be incorporated explicitly in investment decisions. More information will have to be available to facilitate this decision-making.

Case: 'The polluter pays principle': A statement by the OECD

1. Environmental resources are in general limited and their use in production and consumption activities may lead to their deterioration. When the cost of this deterioration is not adequately taken into account in the price system, the market fails to reflect the scarcity of such resources both at the national and international levels. Public measures are thus necessary to reduce pollution and to reach a better allocation of resources by ensuring that prices of goods depending on the quality and / or quantity of environmental resources reflect more closely their relative scarcity and the economic agents concerned react accordingly.
2. In many circumstances, in order to ensure that the environment is in an acceptable state, the reduction of pollution beyond a certain level will not be practical or even necessary in view of the costs involved.
3. The principle to be used for allocating costs of pollution prevention and control measures to encourage rational use of scarce environmental resources and to avoid distortions in international trade and investment is the so-called 'polluter pays principle'. This principle means that the environment is in an acceptable state. In other words, the cost of these measures should be reflected in the costs of goods and services which cause pollution in production and/or consumption. Such measures should not be accompanied by subsidies that would create significant distortions in international trade and investment.
4. This principle should be an objective of member countries; however, there may be exceptions or special arrangements, particularly for the transitional periods, provided that they do not lead to significant distortions in international trade and investment.

(Source: OECD, 1975.)

Social reporting

In principle, it should be relatively straightforward to extend the traditional accountancy role involving financial statistics and statutory information to provide information about the impacts of an organization's activities on the environment. In the 1970s, for example, there were broad developments of 'social reporting' of issues deemed in the public interest. However, in the United Kingdom since the 1970s, there has not been any major change required by the Companies Acts in the information to be reported (which is of a financial nature). As the use of annual reports for public relations purposes grows, an increase in social reporting can be envisaged. At present, the environmental coverage remains generally brief and lacks real depth; for example, reference may be given to say company car policy on unleaded petrol and product safety, and it is often linked to comments on charitable donations and sponsorship, local community work, and so on. The current reporting, essentially financial, should be extended to explicitly assist environmental accountability in at least two ways by:

● inclusion of additional useful and relevant information;
● dissemination to all an organization's stakeholders.

The Institute of Chartered Accountants in England and Wales (ICAEW) research group, for example, recommends that an organization should report formally in their annual report, or in a separate publication, on:

● the organization's environmental policy;
● the identity of the director with overall responsibility for environmental issues;
● the environmental objectives, which should be expressed in a way that enables performance to be measured against them;
● information on past and present actions, including their costs;
● the main environmental impacts of the organization, and, where possible, their measurement;
● the effects of compliance with regulations and any industry guidelines;
● the significant environmental risks not required to be disclosed as contingent liabilities;
● the key features of any external audit report on the organization's environmental record.

Measurement and the monitoring of improvements are important characteristics. Performance information could include energy consump-

tion, emission quantities, noise levels, recycling activities, waste production and so on. For instance, Business in the Environment (1992) has
published recently a report, *A Measure of Commitment*, as practical
guidelines, with case studies, for measuring environmental performance.
In 1990, Shell was fined £1 million after spilling crude oil into the river
Mersey; no reference is given to this incident in its annual report!

In 1990, British Steel published a separate report entitled *British Steel
and the Environment* in which it outlines some of the actions taken to
improve environmental protection at its UK sites. To accompany its 1992
annual report, The Body Shop produced *The Green Book*, which covers its
environmental achievements. In its introduction, the Board Statement
said,

> This Statement is much more than a list of our recent environmental
> achievements. We wanted to create a solid foundation for the future so that
> we could develop new procedures, set new targets, challenge ourselves to
> constantly improve our performance. That's why, along with pointing out
> what we're doing well, we detail our shortcomings too. We've also had the
> whole thing independently verified.

Their independent environmental verification, which was completed by
Arthur D. Little highlighted two problem areas for The Body Shop,
specifically the need to evaluate impacts of its off-site waste shipments
and its discharging of liquid waste into the Channel from its Lit-
tlehampton plant. British Airways and Norsk Hydro, the Norwegian
manufacturing group, were the joint winners of the Chartered Association of Certified Accountants' environmental reporting award in its first
year, 1991. Both organizations have announced that they will undertake
annual environmental audits to assess the impacts of their business
activities on the environment. In addition, Norsk Hydro intends to
include an environmental statement in the financial accounts of its UK
subsidiary. British Airways will publish its environmental assessment
with regard to congestion, noise, waste and staff involvement in the
community. The 1992 award was won by BT.

It will be necessary to avoid potential dangers of abusing the term and
process of 'environmental audit' (see also chapter six). As Andrew Jack
(1992) argues,

> Without compulsory disclosure of environmental information, companies
> are likely to receive growing criticism from outside observers. Without
> independent, objective scrutiny from environmental auditors, the little
> information which is provided will remain suspect.

Case: British Gas plc: Environmental audit

In the autumn of 1990, British Gas initiated a comprehensive environmental audit of the company's world-wide activities with the following objectives:

- to ensure their operations comply with their policy on the environment;
- to identify issues which may require action to improve programmes or management procedures;
- to provide a focus for improved environmental performance;
- to substantiate what the company believes to be a generally good environmental performance.

A pilot phase, completed with environmental consultants, Environmental Resources Limited, demonstrates,

> . . . a high level of compliance with legal and other requirements and few significant environmental problems. However, they identified two key points:
>
> - the need for top-level executive responsibility on environmental matters;
> - the need for more guidance to be given to managers on the interpretation and application of the company's environmental policy statement.

British Gas are now committed to a three-year programme of environmental activity, which commenced in autumn 1992 at an estimated aggregate cost of £5 million. The programme includes:

- further training of in-house auditors;
- the completion of the first cycle of auditing all the organization's 4,000 installations;
- reviews of a number of environmental business issues.
>   (Source: British Gas 1991, Environment Review.)

Financial markets' roles and importance

Many commentators have stressed the power and influence of the financial markets on corporate behaviour, especially with regard to their short-term performance. Given their contemporary significance, the financial markets will have an important role in managing our environment for a number of reasons, including:

- the need to revalue many organizations' assets and liabilities with respect for the environment;
- the need to finance major investments through the capital markets to enhance organizations' environmental performance.

In this section, as well as describing the relatively new phenomenon of 'green' funds, there are both theoretical and practical issues that must be considered, including possible changes in basic accounting principles.

Following the conventional asset/expense distinction, for instance, a cogent argument can be made for organizations being able to capitalize their expenditure for the environment. If permissible, it would decrease short-term pressures on earnings per share (EPS) through amortization over a specified period of time. Using accrual principles, capitalization should be allowed because today's expenditure would provide future benefits. In any event, all new projects should be appraised incorporating explicit criteria of environmental impacts.

In terms of assets, as stressed in chapter one, it is important to differentiate between man-made and natural capital, and also appreciate that some natural capital should be treated specially because it is 'critical'.

There is talk about preserving a 'fully repairing lease' on the world. Valuations of traditional assets and liabilities should be and are being extended. Contingent liabilities, such as a future requirement to clean up and restore a toxic waste site, should be disclosed and provided for. Exxon, for example, spent over $2 billion cleaning up their Alaskan oilspill, and, more generally, it is estimated that American businesses have become liable for more than $100 billion clean up costs for their past pollution of land. Investors and insurers must have a clear comprehension of an organization's future debt burden. Some insurance companies exclude environmental claims from public liability policies, and others have restricted their cover to accidental pollution, rather than the accumulation of pollution over time. In the US, the Securities and Exchange Commission now examine companies' financial statements to determine whether clean up could have a materially adverse effect on their business outlook.

In the United Kingdom, somewhat belatedly, the insurance industry has realized its business exposure to pollution through their policy holders, because public liability policies cover a policy holder against any pollution damage it may cause a third party. For new policies, the Association of British Insurers now recommends explicit exclusion of coverage for the effects of gradual pollution. Liabilities are not confined to business. For example, in a much-publicized case, Leeds City Council forced over five hundred private householders on the Armley estate to pay the £5 million clean up costs to remove from their homes the asbestos contamination that had been caused by a local factory which had been closed for over thirty years.

A number of companies have been forced into liquidation, because their clean up costs far exceeded annual revenues. 'Defendants' (managers) found guilty of environmental crimes have now received prison sentences! Liability must be strict, irrespective of fault or negligence. Moreover:

Should clean up cost be allocated by companies' pollution and waste contribution, or by their ability to pay?

Should liabilities be retroactive, even if past pollution and waste disposals were not illegal at the time?

Under the UK's Environmental Protection Act, local authorities are devising property registers to include any property that could have been contaminated through past or present activities, not only nuclear sites but a full range including domestic dry cleaners. Once registered, unlike in say Denmark, the current intention is that the property remains registered, in part because of the practical difficulty of specifying what constitutes clean land. Banks and building societies, which have lent money to purchase property, possibly on earlier higher valuations, could inherit new liabilities.

In relation to the launch of their privatization prospectus, the UK company PowerGen stated,

Changes in environmental regulations are likely to involve substantial additional capital and operating costs, and may result in some plant closures.

Interestingly, in the UK privatization process, both PowerGen and National Power had plans to invest over £1 billion to satisfy existing anti-pollution targets. However, they wanted assurances that the Government or their customers would pay for any additional investments required because of future EC legislation. Moreover, both companies also desired an ability to be allowed to break environmental legislation without penalty on a temporary basis if one of their major plants breaks down and it is necessary to adopt an increased sulphur burn in other plants.

'Green funds'

In the United Kingdom in the mid- and late 1980s, so-called 'green' funds were launched that focused on environmental principles and performance, as well as the profitability, of individual companies. Ethical investment approaches include both 'hands off' approaches, avoid the defence industry, tobacco, South Africa, and so on, and 'hands on' approaches. Investment products cover unit trusts, pension funds, personal equity plans (PEPs), investment trusts and venture capital funds (see Table 3.3 for a list of some of the unit trusts).

With other 'ethical' funds, the environmental funds represent currently approximately one per cent of the aggregate unit trust market. At present, their significance is more related to the public attention that they are being given by potential investors, rather than simply the scale of their value. Fund managers are now giving more attention to environmental issues in their investment selections. For example, James Capel, the British stockbroker, has developed a Green Index to monitor the business performance of 'green' companies. Different criteria for inclusion are used by different fund managers, although companies should appreciate that the fund managers can decide to disinvest if they believe its environmental performance has deteriorated.

The objectives of this explicit environmental orientation include investments in:

- the 'environmental' industry, such as pollution control equipment and waste management;
- companies whose actions show a real strategic and successful commitment to the environment.

From their start in the private investment area in the United Kingdom, 'green' funds are beginning to attract the interest of the large, institutional investors which represent the main players. However, the success of green funds will be ultimately dependent on the extent that financial advisers raise ethical questions with potential investors and the confidence provided by independent, specialist screening. In the United States, institutions are already active, with a number having signed a corporate code of good environmental practice using the Valdez Principles. The extensive Valdez Principles were established by the Coalition for Environmentally Responsible Economies (CERES) through the significant assets of $150 billion of the Social Investment Forum in the United States. The Valdez Principles are:

- protection of the biosphere;
- sustainable use of natural resources;
- reduction and disposal of waste;
- wise use of energy;
- risk (to employees and communities) reduction;
- marketing of safe products and services;

Table 3.3 Some United Kingdom Ethical and Environmental Unit Trusts

Product	Managers	Launch date	Value (£ million)
Acorn Ethical Unit Trust	Acorn Unit Trust Managers	November, 1988	0.62
Environ Trust	Cooperative Insurance Society Unit Managers	May, 1990	n/a
Fellowship Trust	Credit Suisse Buckmaster and Moore	July, 1986	5.39
Health Fund	Medical Investments Limited	June, 1985	0.8
Merlin Ecology Fund	Merlin Jupiter Unit Trust Management Limited	March, 1988	9.22
Target Global Opportunities	Trust Target Trust Managers	May, 1990	8.6
Amity Fund	All Churches Investment Management	February, 1988	6.5
Conscience Fund	NM Investment Management	September, 1987	9.5
Environmental Investor Fund	TSB Unit Trusts	July, 1989	14.3
Environmental Opportunities Trust	Eagle Star Unit Trust Managers	April, 1989	6
Ethical Trust	Abbey Unit Trust Managers	September, 1987	6.3
Ethical Unit Trust	Scottish Equitable Fund Managers	April, 1989	3.15
Evergreen Trust	Clerical Medical Unit Trust Managers	January, 1990	6
Fidelity Famous Names Trust	Fidelity Investment Services Limited	June, 1985	13
Sovereign Ethical Fund	Sovereign Unit Trust Managers Limited	May, 1989	2.7
Stewardship Unit Trust	Friends Provident Unit Trust Managers	June, 1984	98.2
North American Stewardship Trust	Friends Provident Unit Trust Managers	October, 1985	3.2
Stewardship Income Trust	Friends Provident Unit Trust Managers	October, 1987	15.21

(Source: Simpson, 1991, page 10.)

- damage compensation;
- disclosure of incidents;
- environmental directors and managers;
- assessment and annual audit.

In 1989, the Green Alliance launched these principles in the UK, although, clearly, there is a need to become more specific in operationalizing these rather general, and potentially cosmetic, principles.

It is too simplistic to attempt to describe the specific performance of investments in 'green' funds relative to the general, financial performance of the stock market. However, as Carlos Joly's (1992, page 135) empirical analyses highlight, some general patterns and investment corollaries are possible:

- bear market bottoms are the best time to buy green shares, which is true of most shares and, though not a startling discovery, nonetheless reminds us that the best timing requires the most courage;
- green shares tend to outperform in sustained bull markets, which is true and significant;
- green shares tend to underperform in bear markets, which reflects investor sell-off of high-growth stocks in times of uncertainty;
- and last, but perhaps most significantly in the long run, relative to other equity sectors green shares are extremely attractive for the investor interested in long-term capital gains.

In looking at the financial markets, it is important to appreciate that historical analyses do not necessarily say anything meaningful about the future. Indeed, all advertisements are obliged by law to carry some kind of health warning that prices can go both up and down. However, in the long-run, say a seven to ten years perspective taking us into the next millennium, a cogent argument can be made that, organizations, Government and the general public will value the environment more, and, therefore, 'green' funds are likely to perform well on the stock market.

Government policy

In some sense, in a free, competitive market, it is appropriate for government policy action to attempt to ameliorate disadvantages through a sense of compassion and justice. If the argument for a reliance on the market is founded on efficient resource utilization, it begs the question why it is not helping environmental management today. Simply stated, markets are only efficient and effective, if there is an appropriate pricing policy reflecting full costs and available information for decision-making. Unnecessary wastefulness has not been discouraged, and investments in technical innovations for efficiency and for cleaner operations have not been encouraged.

In many respects, it is incorrect to argue that today's situation represents the outcome of a market failure. Instead, the problems are a policy failure, because it has not been necessary to incorporate all the costs and reflect them in the prices of products and services. The fundamental micro-economic principle of the level of demand being related to price also operates with environmental resources – the lower the price, the higher the demand. Scarcity of environmental resources is not being captured through price; indeed, many natural resources are thought to be free.

The approaches for business to enhance their management of our environment are founded on the need to internalize the full environmental costs of their activities and on the Government's policy backcloth. The Business Council for Sustainable Development differentiates between three approaches:

- economic instruments;
- command-and-control;
- self-regulation.

Any consideration of the relative merits of these alternative approaches should appreciate that the most appropriate way forward is probably based on a mixture of them. Moreover, while the basic objective is environmental management, the administrative costs of implementation and monitoring should not be neglected.

Economic instruments

Economic instruments involve Government intervention in the market operation to affect behaviour, whether it is by taxes, subsidies or tradeable permits. While taxes, such as ones introduced to promote greater energy efficiency, can have a regressive impact because poorer and elderly people spend a higher proportion of their income on energy, the higher price should help to change behaviour and the extra revenue could be used to fund additional welfare schemes to assist the poor. Thus, for effective, long-term management of our environment, as many economic instruments raise revenue for Governments, for their broad support, it is increasingly more urgent for the general public to see explicitly how these revenues are 'recycled' for the benefit of the environment. That is, they should be identified as fiscally neutral instruments.

It is interesting to note that the effects of establishing targets are not necessarily similar to the ones resulting from an introduction of taxes. If the price is specified, the market determines the quantity; if a level is specified, the market sets the price to achieve the quota. Taxes are used to establish targets in terms of prices. If quantities or standards are set, this policy instrument is often called a 'tradeable permit', a kind of licence. For instance, an overall emission level is specified with quotas apportioned to individual organizations; if a particular organization invests in order to reduce its emissions below the required level, it could market its unused element of the quota to other organizations. Thus, this instrument can not only satisfy regulatory requirements, but also offer the attraction of some behaviour choice. There are obviously a number of important practical issues, including the enforcement of permissions, especially by whom? and how? The issue of what happens if an organization pollutes without a permit must also be considered.

In practice, permits have not been introduced for environmental management in the United Kingdom, although they are used in say EC milk quotas. In the United States, the apparently good idea of pollution trade permits has found bureaucratic and political problems. Under the 1990 Clean Air Act, US power stations must reduce their aggregate annual total of sulphur dioxide emissions to nine million tons, a sharp reduction to approximately half the 1980 levels. A command-and-control approach would have established a standard that each power station should satisfy, even though their abilities to meet the target would vary enormously because of the different ages, fuels and technologies of the power stations. For the Act, numbers of one ton allowances will be issued to power stations, the initial allocation being made on the basis of power stations' relative, existing emissions. The issued allowances will be able to be bought and sold in their market. However, while acceptable to private electricity companies, the federal system seems to be unacceptable to a number of public utility commissions. Some coal mining states, such as Ohio and West Virginia are concerned about long-term, local employment of miners. Moreover, with less than 3 per cent of the total allowances available for new power stations, there are obvious concerns about inequities, especially amongst the faster growing southern and western states. The Chicago Board of Trade and the US Environmental Protection Agency are creating a market through their sale of pollution rights, but it may be some years before this market parallels the activities of the credit markets.

On a global level, tradeable permits have been suggested that would permit specified emission levels of particular gases. Based on some kind of auction, these permits could be then traded by countries depending on their expectations of requirements. While there may be start-up difficulties to promote trade and always the possibility of the powerful organizations or countries building up their stock of permits, different people have argued that the system would be much cheaper to operate than any one founded on regulation (see, for example, Tietenberg (1990)).

Command-and-control

The command-and-control approach is for government regulations to specify standards to satisfy specific environmental objectives. 'Performance' standards are targets that organizations can achieve through different, self-selected ways; 'prescriptive' standards are ones for which the organizations have to use specific technologies.

Regulations will remain a basic approach to improving environmental management, but, in practice, they can be inflexible, static, and relatively costly to administer. However, as illustrated in Germany where environmental regulations are generally stricter than in most other countries, they can help to provide an impetus for innovation and competive business advantage (see, for example, Michael Porter (1990)).

Self-regulation

While, unfortunately, too many organizations believe that mere compliance with existing regulations is sufficient, self-regulation can offer more flexibility and opportunities for innovation, as well as permitting an organization's senior management to drive its priorities and actions. Growing corporate social responsibility can be a basic driving force behind self-regulation (see also chapter seven), but so can the threat of future stricter Government regulations and increased pressure from the organization's stakeholders.

In the long-term, this approach may be the most efficient and effective, because the organizations have direct access to information for their decision-making. However, as indicated in the earlier chapters, there has to be a business benefit through the operation of market forces which reflect full environmental costs, otherwise some organizations may be able to get competitive advantage by not operating on a 'level playing field' through a disregard of their detrimental impacts on the environment.

Case: 'King Coal': What is the real value of scarce finite resources?

In October, 1992, British Coal, with the support of the British Government, announced the closure of thirty-one coal mines, representing over half the country's coal industry. While extreme pressures from politicians of all parties and the general public has resulted in a decision to further investigate the potential options, the proposed action to close existing coal mines before all their coal has been mined raises important economic, as well as political and social, issues. Remember, the closure of a coal mine is forever!

Lord Justice Glidewell ruled that both the Government and British Coal had acted,

... unlawfully and irrationally

in its ignorance of miners' rights and its lack of consultation with the unions.

In this brief discussion, the following interrelated perspectives are highlighted:

● economic;
● regulatory;
● social;
● geological.

From an economic perspective, many professionals state that the market forces argument behind the closure of pits is flawed. British energy markets have a number of distortions, in part because recent privatisations of state-owned enterprises have created large (regulated) monopolies and quasi-monopolies. The fundamental question is: On what basis is a mine uneconomic?

The significant relationship is between the realizable price for coal and the ongoing costs of operating the mines, rather than any 'booked' accounting costs of coal production including capital charges. If the decision is about closure, sunk capital and depreciation costs, become irrelevant.

From a regulatory perspective, Professor Stephen Littlechild, who is responsible for regulating Britain's electricity industry, has argued that competition will take many years to develop, especially in generating, with some major organizations retaining a large market share. More specifically, with regard to the pool pricing system, Littlechild believes that the current control system means regional electricity companies have little incentive to control costs because they are permitted to pass on any increased costs directly.

> Problems have arisen in the pool, in that the major generators have been able to influence prices . . . in general, the pool should facilitate competition and improve options open to customers, generators and suppliers, rather than impose a straitjacket on the development of the market.
>
> (Littlechild (1992) Institutional Investor, quoted in *The Times*.)

In December, 1992, The Trades Union Congress complained to the European Commission, to Sir Leon Brittan the Competition Commis-Osioner, that the competition between coal and gas is unfair. It is argued that the privatized electricity companies' shareholding in gas-fired power stations breaks the Treaty of Rome.

From a social perspective, it is sufficient to stress that the Government did not expect the scale or scope of the national outrage at their proposed closure of coal mines. The direct impacts on miners, their families and their communities are thought by people from all walks of life to be morally indefensible and socially unacceptable.

From a geological perspective, there is not only the premature abandonment of mines and the loss of their reserves, but also the medium- to long- term issues of further reliance on other finite fossil fuels. Any decision to close a mine that possesses significant reserves must be taken after very detailed consideration. The current pithead stocks of coal, in part the result of decisions taken after the adverse impacts of coal shortages arising from the miners' strike in the 1970s, have also been the result of the decreasing coal-based, and increasing gas- and nuclear-based, electricity generation. Indeed, arguments can be made to suggest that natural gas cannot be really justified as the fuel to generate electricity. For instance:

- natural gas is an important chemical feedstock in its own right;
- natural gas is itself a versatile fuel that can be stored and distributed;
- reserves of natural gas are short-term, lasting only decades, while the reserves of coal could last for centuries.

A future founded on imported coal, gas, . . . at unknown costs, has been highlighted as a national risk by a number of commentators with potentially adverse impacts on the country's balance of payments and geopolitical risks associated with their continued supply. Enhanced management of scarce resources, however, is only likely with a single, national Energy Policy that provides a co-ordinated evaluation of the resource base, developing technologies, costs and of appropriate pricing.

Anyway, rather than British Coal and the Department of Trade and Industry deciding to close specific mines, it may be appropriate to leave them to the market, allowing individuals/organizations to bid for them. Can another organization reduce operating costs to compete with imported, often subsidized, coal and natural gas? At least, the sale of some mines could mean that British Coal, which is still to be privatized, has some direct competitors!

Postscript
After further review, four independent reports agreed that British Coal had made a reasonable estimate of the future demand for coal and of the pits to be closed. Thus, the problem became a political one for the Government to decide the extent to which they intervene in the coal market. A reprieve for thirteen mines was announced in March, 1993, costing £500 million, although it is probably more likely to be a short-term palliative than a long-term solution.

The role of Government is not confined to national, domestic policies. The environment and its problems cross national boundaries, and many major initiatives are of an international nature, such as the Rio de Janeiro Earth Summit and EC Directives. The Montreal Protocol on the ozone layer is an example of international co-operation and regulation, with many countries having their own emission standards. When global and/ or national standards or targets are established, significant distributional effects can be envisaged. Transitional arrangements may have to be negotiated. For instance, the ten year period of grace and less restrictive production and consumption targets for developing countries to achieve their target for reduced chlorofluorocarbons (CFCs) consumption under the 1987 Montreal Protocol and the subsequent 1988 Helsinki and 1992 Copenhagen Agreements, with the last one agreeing to bring forward the CFC phase-out deadline from 1 January, 2000 to 1 January, 1996. The means by which individual countries are attempting to achieve their targets vary, some combination of the market mechanism and/or regulations.

Inequalities, whether in education, health, position, wealth, . . ., exist in most human societies. It is appropriate here to reference some basic concepts of justice and responsibility (but see chapter seven for a longer discussion).

To some, the global environmental crisis is primarily a crisis of values. In this view, the basic cause of the problem is that we as a civilisation base our decisions about how to relate to the environment on premises that are fundamentally unethical.

(Gore, 1992, page 242.)

More specifically;

In today's world, the links between social injustice and environmental degradation can be seen everywhere: the placement of toxic waste dumps in poor neighbourhoods, the devastation of indigenous peoples and the extinction of their cultures when the rain forests are destroyed, disproportionate levels of lead and toxic air pollution in inner-city ghettos, the corruption of many government officials by people who seek to profit from the unsustainable exploitation of resources.

(Gore, 1992, page 247.)

Case: Tropical forests: A policy jungle

Much-needed income for some poor, developing countries comes both directly from commercial logging and indirectly from agricultural crops grown on land that was previously forest. For the Earth, however, forests are extremely important, because they are a major home of biodiversity and they represent a significant component in the carbon cycle.

It is evident that many developing countries resent the apparently new morality of developed countries, many of whom have already cleared their own forests. For instance, arguing that it contravened GATT rules, the Association of South East Asian Nations (ASEAN) recently criticized Austria's introduction of a mandatory tropical labelling scheme as *unilateral and discriminatory*. It is noted that Austria's domestic temperate timber industry is not insignificant.

Ed Barbier discusses the debate over tropical deforestation by highlighting the spectrum between 'preservation' and 'development'.

Why are market forces creating the current situation?

As David Pearce (1991, pages 27–28) stresses,

. . . a great deal of deforestation would be avoided if markets were managed to function more efficiently. It is because of direct government interference that the price signals are distorted, making timber extraction – and, more importantly, clearance for agriculture – profitable.

What the analysis reveals is that the world is needlessly destroying much of its most precious ecosystems – the rainforest – and often not in the name of profits for people who need those profits to survive. If that is right, the policy measures are clear and, in terms of economic cost, relatively cheap. That does not make these policies easy to implement, because of the conflict of interests they generate, but we cannot pretend that conserving many of the world's ecosystems is expensive. It is not a matter of expense so much as one of correcting wrong incentives, and in changing incentives to protect ecosystems we will conserve biodiversity.

Barbier indicates that the problems of tropical deforestation are related directly to countries' economic and non-economic policy distortions, the latter including land ownership and regulation. Interestingly, Al Gore (1992, pages 180–181) argues that,

... in many countries, corruption is one of the principal causes of environmental destruction. To take only one of literally thousands of examples, concessions to clear-cut the rain forest of Sarawak, in East Malaysia, were sold personally by the minister of environment for Sarawak. Even though he was officially responsible for protecting the integrity of the environment, he enriched himself personally by selling permission to destroy it.

Over half of Sarawak's forest is categorized as 'permanent forest estate' for timber production. In 1990, experts for the International Tropical Timber Organization reviewed Sarawak's forest position, recommending the reduction in annual log production from this estate to ensure sustainable future harvests. Given this advice, Sarawak is targeting a reduction of a quarter to be phased in over two years. It should not be underestimated that, for the timber companies, the scale and speed of this change raises a number of fundamental business and management issues regarding the new quotas.

Concluding comments

In the 1980s, the orientation of accounting and finance was towards an emphasis on financial efficiency and value for money. With the growing interest in the environment, along with the implications of recent financial scandals and business failures, there are signs that greater attention is being given to accounting and finance in the public policy arena. The extent to which the environment poses fundamental challenges to the market-based system in which accounting is placed remains a basic issue. Some people argue that improvements within the existing structure can be forthcoming, while other people argue for a more radical transformation of accounting to one that is not preoccupied with economic growth. Any developments must be within the overall context of continued wealth creation.

It is reassuring to know that Finance Directors are beginning to appreciate the significance of the environment, particularly because of risk exposures, market opportunities and general social responsibility. A 1990 Coopers & Lybrand Deloitte survey, for example, indicated that 87 per cent of the Finance Directors of major British companies view the environment as a business issue for their organization, and 55 per cent think it raises significant financial concerns.

It is essential that we derive measurements to monitor progress against sustainable goals, rather than use existing statistics to measure developments against inappropriate goals. It is incorrect and misleading to argue that, if you cannot measure it, you cannot manage it! Computer-based information systems and the availability of data are becoming more important (see chapter six for a more extensive discussion, including a consideration of environmental auditing).

Unfortunately, being realistic, with today's scale of consumer, corporate and national debt, the confidence and stability desired by the financial markets is not present; unfortunately, the continuation of this situation is likely to be a significant constraint on investments for sustainable development. Moreover, as indicated by this chapter, many of the basic foundations need to be revisited for managing our environment. Until the appropriate professional bodies for accountancy begin to take these issues beyond the discussion stage, no real change in philosophy or progress can be envisaged. For instance, how should companies account for the 'free' environmental resources that they use?

Review questions

1. How should an organization measure its investment in managing our environment?
2. Comment on the strengths and weaknesses of conventional national income statistics, particularly with regard to their measurement of environmental importance.
3. Discuss some of the difficulties in completing an investment appraisal of an environmental project.

Study questions

1. To what extent do you believe that there are only three issues in the business management of environmental investments: managing costs; managing benefits; and managing risk exposure?

2. Using secondary data, such as annual reports, industry surveys and other documents, develop a picture of the scale and nature of investments for our environment across different industries and different organizations.
3. Consider the future of 'green' investment funds.

Further reading

Anderson, Victor. (1991) *Alternative Economic Indicators*, Routledge, London.

In so-called 'New Economics', the intention is to extend traditional economic debate by incorporating environmental and human factors that are usually omitted from economic theory. In a practical way, Victor Anderson considers what needs to be measured, and proposes various statistics that can assist analyses. Criteria are suggested as the characteristics of 'good indicators':

- the underlying data should be already available or easily and cheaply collectable;
- an indicator should be easily understood;
- an indicator must be about something which is measurable;
- an indicator should measure something which is significant in its own right or reflect something which is significant;
- there should be a short time lag between the issue of interest and the associated indicator becoming available;
- it is useful if an indicator can be disaggregated by characteristics of interest;
- international comparability is desirable.

Victor Anderson then proposes five, priority environmental indicators:

- deforestation in square kilometres per year;
- carbon dioxide emissions from fossil fuel use, in millions of metric tons per year;
- average annual percentage rate of increase in population;
- number of operable nuclear reactors;
- energy consumption (in tons of oil equivalent) per million dollars of GDP.

Business in the Environment (1992) *A Measure of Commitment*, BiE, London.

Developed with KPMG consultants, this clear and comprehensive document provides practical guidelines for measuring an organization's environmental performance (see also chapter six).

To illustrate different aspects of performance measurement, the following case studies are presented:

- Bluecrest Convenience Foods Limited: effluent management within a process company;
- British Airways plc: measures of corporate environmental performance;
- BT plc: supplier performance management within a major procurer;
- Ford Motor Company Limited: recyclability of motor vehicles;
- Hydro Aluminium Metals Limited: energy management in a process environment;
- IBM United Kingdom Limited: a system to assess the environmental impact of products;
- Kodak Limited: measuring the effectiveness of environmental communication and training;
- John Laing plc: energy efficiency in temporary site accommodation;
- National Power plc: measuring effectiveness of environmental management systems in a power company;
- National Westminster Bank plc: energy management and waste segregation, disposal and recycling in a large service organization;
- Safeway Stores plc: measuring store waste in the retail sector;
- Scott Limited: development of a system to assess and set standards for pulp suppliers;
- Sutcliffe Catering Group Limited: measuring supplier packaging and transport performance within a catering company;
- Wessex Water plc: measuring energy savings in process electricity consumption.

Cairncross, Frances (1991) *Costing the Earth*, Economist Books, London.
If people in business or in Government had to choose one management book on the subject of the environment, a strong case can be made for this one. It is highly lucid, with firm arguments supported by many examples and pragmatic suggestions. Frances Cairncross sub-titles her book:

- what Governments must do;
- what consumers need to know;
- how businesses can profit.

At the end, a checklist for companies of a dozen suggestions is provided:

- put the most senior person possible in charge of environmental policy;

- draft a policy and make it public;
- measure;
- institute a regular audit to check on what is happening;
- communicate;
- consider ways to reduce the range of materials you use that could do environmental harm;
- think about the materials in your product;
- remember that you may be able to make a business opportunity out of disposing of your product when the customer has finished with it;
- if you invest in a country where environmental standards are low, do not expect them to stay that way;
- accept that green regulations will tend to converge upwards;
- be flexible;
- remember that greenery is often a proxy for quality in the eyes of your customers, your workers and your managers.

Feshbach, Murray and Friendly, Alfred (1992) *Ecocide in the USSR*, Basic Books, New York.

Since the euphoria of the fall of the Berlin Wall, many commentators have highlighted the previously unknown, enormous damage to the environment from the communist regimes of Eastern Europe. The horrendous examples that have been used to illustrate the situation are presented by Murray Feshbach and Alfred Friendly in a clear and well-argued book.

The basic factor behind today's problems in the former Soviet Union is suggested as the idealism of a central planning system that was founded on single aggregate output measures with the view that natural resources are free goods.

Gray, Robert (1990) *The Greening of Accountancy*, The Chartered Association of Certified Accountants, London.

A broad, albeit superficial, descriptive discussion of some of the issues. Important because the report feeds into a professional body.

Owen, Dave (ed) (1992) *Green Reporting*, Chapman & Hall, London.

While suggesting that the environment will be the challenge of the nineties for the accountancy profession, this expensive volume is disappointingly uneven in quality and coverage. As the title indicates the emphasis is on reporting, and, while a number of different actors are covered, the specific discussion on the trade union movement merits consideration.

Pearce, David W. (1993) *Economic Values and the Natural World*, Earthscan, London.

A slim volume that clearly describes the way in which economists try to measure preferences, whether for improvements in the natural resource asset base or in environmental quality or against environmental degradation. Putting monetary values on environmental preferences is a non-trivial task. Given the long-standing interest of economists in cost-benefit analysis and associated public policy, it is helpful to the reader that David Pearce not only considers the why? and the how? in an environmental context, but also examines what can be done with the analyses. A range of different cases offer useful examples of economic valuation studies in practice.

4 Marketing management

Whereas consumerists focus on whether companies are efficiently serving consumer material wants, environmentalists focus on the costs imposed on the environment in serving these needs and wants. . . . Environmentalists are not against marketing and consumption; they simply want them to operate on more ecological principles. They think the goal of the marketing system should be to maximize life quality. And life quality means not only the quantity and quality of consumer goods and services but also the quality of the environment.

(Philip Kotler, 1988, page 152.)

Introduction

Marketing is a recognized business function undertaken in all organizations. It is being increasingly appreciated by many organizations that marketing has a significant managerial and cultural orientation. While Theodore Levitt's (1960) paper on 'marketing myopia' highlighted the need to move away from a production/sales approach to a business approach, there is growing recognition that everybody in an organization should be concerned with marketing, seeing it as a corporate activity involving not only the production of products and services but also their distribution and ultimate consumption.

In this chapter, we discuss the traditional marketing concepts, principles and practices and their implications for managing our environment. Important changes in public opinion, the 'green' consumer concept, arising from greater awareness and concern for the environment, are also considered. The argument is developed to demonstrate that the marketing function and its activities need to be incorporated into an organization's corporate social responsibility philosophy (see also chapter seven). There is a need for organizations to move from merely a superficially green PR tinge towards a deeper environmental consciousness which should be beneficial to organizations' stakeholders in the long-term.

There is no doubt that changes in consumers' expectations and tastes are having important impacts on manufacturers and retailers of consumer goods. While the focus of this chapter is on the increasing significance of the environment for consumer marketing, industrial marketing should not be neglected, and some attention is given to this topic (see also chapter five). To question marketing's conventional wisdom, after a brief consideration of marketing strategy in the next section, the bulk of the chapter is devoted to an examination of the so-called marketing mix. The discussion is then broadened to other issues, particularly consumer sovereignty and business ethics.

Conventional wisdom within the marketing field in Western capitalist and democratic societies has been closely linked to consumption. Indeed, it can be argued that the political changes in Eastern Europe have been driven less by a drive for democracy and more by a drive for consumerism. The promotion of consumerism and the creation of new demands has led to our Western society being labelled 'disposable' or 'affluent', which is generally thought to be incompatible with increased environmental concern.

How can consumers' environmental concerns affect purchasing decisions and consumption patterns?

The definition of marketing as profitably satisfying customers' wants and needs emphasizes that marketing no longer is about products and services *per se*; the focus of attention must be on consumers (or potential consumers) and how to provide products and services in order to satisfy and, indeed, develop, their demands in the context of the overall market. Moreover, the mass consumption of the decades following the Second World War is being replaced by growing concerns for service and quality with the development of a mosaic of desired individual and niche consumption patterns.

This perspective of the 'customer as king', in the sense that he/she should start and finish the marketing process, is potentially significant. In part, it can explain the 'market pull' by environmentally conscious consumers, forcing businesses to react to the increasing number of green demands, among other things by changing both the nature by which businesses market and sell themselves and their products and services.

Empirical evidence from the 1990 MORIScope of Britain, for example, indicates the strength of public concern with only law and order and the health service deemed more serious, and the environment being much more important than both inflation and unemployment. Seventy-five per cent of the MORIScope respondents believe that Britain should,

> ... emphasize protection of the environment at the expense of economic growth;

and only twelve per cent believe that Britain should,

> ... emphasize economic growth at the expense of the environment.

At an international scale, a recent *Health of the Planet Survey* of twenty-two countries by the George H. Gallup International Institute shows a high level of commitment to environmental protection at the expense of economic growth. Respondents were asked to choose with which one of the following two statements they agreed with most,

> 'Protecting the environment should be given priority, even at the risk of slowing down economic growth'
> Economic growth should be given priority, even if the environment suffers to some extent.

Table 4.1 summarizes the responses across a range of countries at different 'development' stages.

It could be argued that truly environmentally conscious consumers should consume less, not only differently. The basic question, however, is whether it is a realistic scenario. People are demanding; generally, they want both a clean environment and an expanding range of products and services.

Changing lifestyles should not necessarily mean a lower quality of life. The carrying capacity of the environment must be viewed as a determining factor if we want to maintain, or hopefully increase, the quality of life for people. In a thoughtful, albeit somewhat idealistic, book, Gregory Bateson (1988) examines the connections between *Mind and Nature*, arguing for 'a necessary unity' to preserve both ourselves and our planet. It is becoming increasingly apparent that we cannot continue either the excessive scale or pattern of today's consumption, which is detrimental to our planet and its people, and, in *Our Common Future*, there is the suggested objective of,

> ... lifestyles within the planet's ecological means.

Table 4.1   Economic Growth versus
the Environment

Country	Chose Protecting the Environment over Economic Growth (%)
India	43
Philippines	59
Turkey	43
Chile	63
Poland	58
Mexico	72
Brazil	71
Hungary	53
Uruguay	64
Russia	56
Republic of Korea	63
Ireland	65
Great Britain	56
Netherlands	58
Canada	68
West Germany	73
Denmark	77
United States	59
Finland	72
Norway	72
Japan	58
Switzerland	62

(Source: Gallup, 1992.)

Marketing strategy

Marketing strategy and planning is concerned with identifying and servicing customers for long-term profit. Igor Ansoff's (1987) well-known strategic matrix provides a framework to consider marketing objectives in terms of products/services and markets. Simply stated, as shown in Figure 4.1, marketing objectives are concerned with combinations of existing and new products or services and markets, that is:

- existing products/services in existing markets;
- existing products/services in new markets;

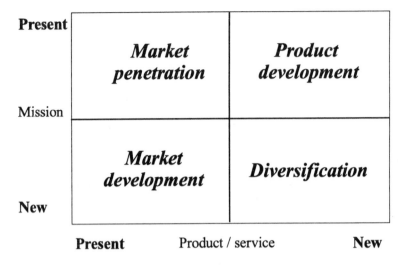

Figure 4.1   Ansoff's Strategic Matrix

- new products/services in existing markets;
- new products/services in new markets.

Martin Charter (1992, page 165) uses this matrix to specify 'greener marketing' objectives and strategies. That is, a simple extension gives the objectives as:

- existing greener products in existing greener markets;
- new greener products in existing greener markets;
- existing greener products in new greener markets;
- new greener products in new greener markets.

With regard to the last objective, after outboard motors were banned from Swiss lakes owing to concerns about oil pollution, Castrol developed Biolube 100, which gave 65 per cent less hydrocarbon emissions, for this small, but lucrative market. Figure 4.2 is a corrected version of Martin Charter's adaptation of Ansoff's matrix.

Once specified, these objectives are addressed by marketing strategies which are founded traditionally on the so-called 'marketing mix'. Georg Winter (1988), who developed an 'integrated system of environmentalist business management' which is supported by the Commission of the European Communities, includes a checklist around this traditional marketing mix as a kind of practical handbook. As chapter two highlights, there are a range of different strategic directions that

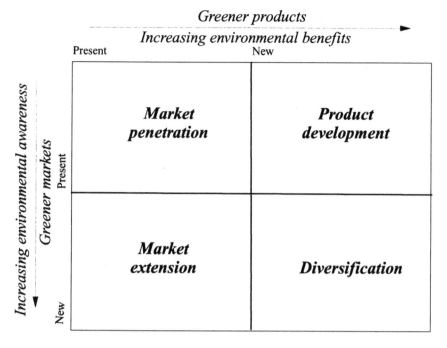

Figure 4.2   Greener Marketing Objectives
(Source: Charter, 1992, page 165.)

organizations can take. Interestingly, Martin Charter (1992, page 161) describes Ken Peattie's eight types of responses to greener marketing opportunities:

● *Head in the sand*   The Quality Director of Addis was quoted as saying that the heavy metal cadmium is 'safe' and that,

> There is no legal requirements to change yet, so I don't see why I should.

Heinz made major donations to the World Wide Fund for Nature, and sponsored Green Consumer Week, but were forced to withdraw from the event to minimize bad publicity over their subsidiary Starkist, a company involved in the destruction of dolphins through the use of drift nets for tuna fishing.

● *Defensive*   The British Aerosol Manufacturers Association, ISC Chemicals and ICI complained to the Advertising Standards Authority over Friends of the Earth's campaign on CFCs, stating that,

> ... the conditions of the ozone layer over the Antarctic ... were unrelated to the use of CFCs in aerosols.

Defensive strategies also came following the publication of *The Green Consumer Guide* and *The Good Wood Guide*. Companies that were criticized tended to respond by disputing or attempting to disprove the evidence in the guides. However, some retailers took a more proactive response and switched timber sourcing to managed plantations, sponsored tropical reforestation research and donated to Britain's Woodland Trust.

● *Lip service*　The Soap and Detergent Industry Association admitted that,

> Many misleading claims are being made about the environmentally friendly nature of various Green products in order to attract the buying power of the Green Consumer .... We are concerned that goodwill will be frittered away chasing illusory benefits claimed for some Green detergents.

This followed the rush of confusing and meaningless claims made about 'phosphate-free' washing up liquids (since none contain phosphates anyway), and the 'biodegradability' of some detergents. Another approach has been to donate to charitable causes connected with the environment while leaving products unchanged, or only slightly environmentally friendlier. This can backfire if the attention of green pressure groups is attracted.

● *Knee-jerk*　Some companies have reacted to environmental pressures with actions rather than words, but with actions which are unplanned and aimed at defusing a particular potential environmental problem. Habitat reacted almost immediately to Green Consumer Week by announcing that they were abandoning the use of tropical hardwoods in their furniture, to the surprise of many managers.

● *Piecemeal*　Shell has launched some very positive environmental intentions, such as the 'Better Britain Campaign', but it continues to manufacture some of the world's most destructive chemicals.

● *Green-selling*　A number of companies have altered their products and switched their sales pitch to highlight the new or existing environmental benefits of their products. Aerosol companies have developed a range of propellants to replace hard CFCs, and added labels such as 'ozone friendly', 'ozone friendlier', or 'ozone safe'. What many consumers do not realize is that a label saying 'ozone friendlier' tends to mean the can contains 'soft' CFCs which still damage the ozone layer, but more slowly and to a lesser extent.

● *Integrated greener marketing*　In green-selling, the emphasis is about communicating the benefits of products. The degree of benefit offered to consumers, and the extent to which it matches their needs is not really considered. Developing Integrated Greener Marketing will mean looking at matching the environmental performance of products and production processes with a view to the current and future environmental concerns of consumers and stakeholders. This should lead to a more proactive corporate response, with a consideration of environmental issues being built into each aspect of strategic market planning.

● *Integrated organizational* The desire to respond to the needs of green consumers and tougher environmental legislation will spread to all parts of the greener business. Corporate strategy, investment decisions, purchasing decisions and corporate policies should all be developed with environmental concerns in mind. There are reasonably few examples of organizations which have gone green beyond tactical marketing, but the two best examples are probably 3M and The Body Shop.

Case: Promoting car performance: Motoring or environmental?

Simply stated, the advertising dilemma is to promote a car's 'performance', as well as being environmentally sound (see, for example, Vauxhall's general 'greenlease' advertisement).

At the time of writing, car manufacturers are facing severe problems because of reduced sales, owing, in part, to the global economic recession and also to additional production facilities being opened by new supplier countries. While many of the companies need to strategically reorientate to sell fewer cars profitably, relaxing their past preoccupation with market share, it can be argued that, in the medium to long term, environmental pressures will be the major force behind transformation of both this product and this industry. Some of the changes will be driven by the competing car manufacturers themselves, and others will be driven by consumers. Ford, for example, is the volume market leader in the UK.

> Environmental concerns could lead to a shift in the manner in which the private motor vehicle is viewed and utilized in modern society, which, in turn, will affect consumer purchasing decisions. Tax incentives to make smaller, even more fuel efficient vehicles more popular among consumers would be welcomed by Ford.
>
> (quoted by Rogers, 1991, page 51.)

The 1992 Motor Show, at Birmingham's National Exhibition Centre, highlighted explicitly car manufacturers' environmental efforts, including an Environmental Day. In November 1992, at a Frankfurt symposium on the automobile industry and the environment, representatives from Daimler-Benz, General Motors, Mazda, Volkswagen and Volvo appealed for global industry co-operation to combat environmental problems.

The marketing mix

The traditional marketing mix, which is known commonly as the 'four Ps', involves:

- *Product:*
    - expand the product line;
    - change performance, quality or features;
    - consolidate the line;
    - standardize design;
    - modify positioning;
    - change the mix;
    - enhance branding;
- *Price:*
    - alter price, terms or conditions;
    - introduce skimming policies;
    - consider penetration policies;
- *Place:*
    - change delivery or distribution;
    - change service;
    - change channels;
    - change the degree of forward integration;
- *Promotion:*
    - alter advertising;
    - change selling.

Given the nature of the marketing function, the availability of useful and actionable information is very important, and is becoming even more significant as organizations face more and better competition. Indeed, as indicated in Figure 4.3, the proposal of a fifth P, (data) Processing, that comprises data collection, data analysis and data presentation, seems reasonable (see also the discussion in chapter six).

The environmental challenge should influence all of these five components of the marketing mix. The trend in public opinion towards greater environmental consciousness and concern has been reinforced by the mass media turning 'green' and its direct influence on the formation of people's attitudes and opinions. The environmentally conscious consumers are here to stay and their demands and influences on business are likely to rise even further. In many respects, therefore, and in relation to marketing, the environment can be thought of as the new consumer issue for the 1990s.

By considering each of the five Ps in turn from a more environmentally conscious perspective, it is possible to explore some of the changes that have already taken place within the functional area of marketing, and also the scope for further changes towards 'environmental excellence'.

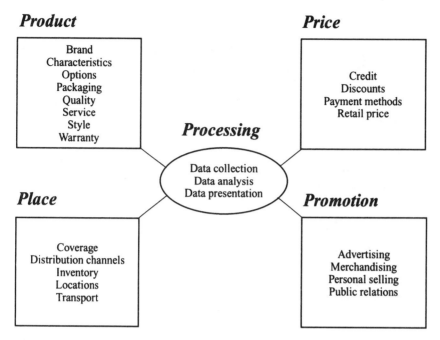

Figure 4.3 The 'Five Ps' of the Marketing Mix.

Product

The fundamental question under this component is:

What products/services should an organization bring to the market?

Traditionally, this question of product decisions has been closely linked to the issue of choosing the right product portfolio, with various products at different stages of their product life cycle (new or existing products). As each stage of their life cycle is affected by different competitive conditions, each stage requires different marketing strategies if sales and profits are to be realized efficiently.

However, when considering the product with regard to the environment, management must adopt a broader perspective of a product's characteristics. Companies must accept that they should not only consider the product's affects on the environment when used or disposed of by the consumers, but also examine the complete life cycle impacts. A 'life cycle' stewardship orientation relates to the product's adverse environmental considerations from design, through manufacture, storage, packaging, use to disposal. This concept of 'cradle-to-grave' management, which is slowly being adopted by some organizations in some industries, is discussed more fully in chapter five on production and operations management. In this sub-section, attention is given to the marketing-related product features of product quality and packaging, as illustrated in Figure 4.3.

A 'conserver' mentality is closely linked to the general discussion about needs and wants, and specific trade-offs between product fashion and product quality, especially durability and reparability. Table 4.2 lists some of the possible options for product improvement.

Table 4.2   Corporate Options for Product Improvement

Eliminate or replace product.
Eliminate or reduce harmful ingredients.
Substitute environmentally preferred materials or processes.
Decrease weight or reduce volume.
Produce concentrated product.
Produce in bulk.
Combine the functions of more than one product.
Produce fewer models or styles.
Redesign for more efficient use.
Increase product life span.
Reduce wasteful packaging.
Improve repairability.
Redesign for consumer reuse.
Remanufacture the product.

(Source: Schmidheiny, 1992, page 110.)

In his book *Greening Business*, John Davis (1991) distinguishes between improvements in quality in products made for industrial use and consumer products, arguing that durability and repairability of the former have improved enormously. However,

> The same thing has not happened with consumer products. On the contrary, durability and repairability have in many cases been reduced under the influence of fashion marketing and the drive for high-volume production. Consequently, the general public no longer considers these particular qualities to be as important as they once did.
>
> It was the substitution of the concept of marketing for the traditional notion of business as the provider of the goods and services essential for the sustenance of a dignified life that fundamentally changed the concept of quality. If one takes, for example, the universal human need for self-respect, there is available a vast potential range of goods that may claim to meet it.

Any single item in the range may have little intrinsic worth in terms of performance or of durability. An item of clothing may provide little warmth and may only survive for a few weeks; another may be thrown away because it cannot be cleaned. The quality of such products would be judged largely on aesthetic or fashion grounds, which are as changeable as the weather, or might be signalled by a so-called designer label. Such is the influence of the marketing concept on the meaning of quality.

(John Davis, 1991, page 147.)

And further,

Repairability has not yet featured significantly as a quality characteristic of consumer goods, though it is an important consideration in the choice of industrial plant and for products used in commerce. With energy efficiency, the importance of repairability will spread once the economic and environmental advantage of extended useful life becomes recognised. In the long run the contribution that both durability and repairability can make to a less wasteful use of scarce and increasingly expensive materials will be valued.

Thus, with this new meaning of quality, a product or service will come to be seen as good not only because of its intrinsic worth to the user, but also because in every respect it contributes to a sustainable way of life. To be of good quality it must not only satisfy all the traditional need of individuals, but must also contribute positively to the general well-being through the sparing use of resource, appropriate technology and environmental harmony.

(John Davis, 1991, pages 151–152.)

In marketing, the quality of a product is often inseparable from the quantity of the packaging.

Packaging is the ultimate symbol of our consumer culture.

(Stilwell and colleagues, 1991, page 1.)

There is little doubt that often consumers are attracted to or deceived by lavish packaging, which symbolizes our so-called affluent, disposable society. It is not only the increasing quantity of packaging, but also the nature of the packaging that makes disposal and recycling difficult. Packaging is a deceptively complex issue, and also a profitable business.

The primary function of packaging is to conveniently contain and to safely and hygienically protect the contents from the time of their manufacture through storage and distribution until consumption or initial use. The secondary function is to give information on the product's quality and performance, and perhaps also to provide a means of brand promotion.

Joseph Stilwell and his colleagues (1991) highlight that environmentalism is a very recent issue in the packaging industry, and they describe the evolution as:

- *1960s:*
  - convenience as a basis of competition: 'throw away society';
  - pop marketing: 'the silent salesman';
- *1970s:*
  - fragility assessment: analytical design for movement;
  - energy shortage: lightweighting (the first source reduction);
- *1980s:*
  - malicious tampering: 'safe' packaging;
  - extended shelf-life: barrier technology; controlled atmosphere technology;
  - quality: sensory issues;
  - industry restructuring: 'evaporation' of resource;
- *1990s:*
  - environmental issues.

From an environmental perspective, packaging can be examined at a number of different stages. Firstly, without reducing the required product protection, manufacturers and retailers should explore whether the quantity of packaging can be reduced at source. Secondly, the extent to which the packaging can be reused should be considered (see, for instance, The Body Shop's refill approach to packaging which has been followed by other organizations). Thirdly, the recyclability of the existing packaging should be evaluated. As well as being helpful to the environment, reduced packaging may provide other business benefits, including lower transport and storage costs. Packaging can also be reduced by product modification, such as selling household cleaning products in a more concentrated form. Procter & Gamble introduced this idea in the United States in late 1989 with an explicit environmental claim; Downy Refill fabric softener comes in milk carton-type containers for mixing with water in a used Downy plastic bottle. In a much smaller package, the refill costs less than the regular Downy. This concept has now spread successfully across Europe.

In Denmark, reuse of beer and soft drinks bottles is stimulated by 1981 legislation that requires their sale only in returnable bottles with a compulsory deposit. Metal cans were banned, and the requirement of reusability effectively means plastic bottles are non-feasible. The success of the scheme is demonstrated by a return rate of over 95 per cent. However, complaints from competing foreign brewers led to a protectionism case being heard by the European Court of Justice. In 1988, the Court supported the Danes, because it felt that environmental considerations are important and that the practice had not been introduced to protect its domestic industries. Later, the European Commission indicated that it would not stop countries introducing national schemes. Germany, for example, as well as introducing a mandatory deposit on plastic bottles, sets targets for the percentage of containers that must be returned:

- 90 per cent of beer and mineral water containers;
- 80 per cent of carbonated soft drink containers;
- 50 per cent of wine bottles;
- 35 per cent of non-carbonated, soft drink containers.

More generally, the Institute of Packaging Professionals has established packaging guidelines, which are classified under the following headings:

- source reduction;
- recycling;
- degradability;
- disposal;
- legislative considerations.

Overpackaging, however, is used to differentiate companies and their products in the marketplace, and, in recent times, packaging has become a very powerful marketing tool. As pointed out recently by Joseph Stilwell and his colleagues (1991), some firms' major product line is packaging and garbage. For instance, it is noted that packaging is the largest single component of municipal solid waste in developed countries. It is, however, unfair to portray a completely negative picture. In terms of the total waste created in a western, industrial economy, packaging represents less than 5 per cent, although it is often the most visible form. Moreover, because of effective packaging, the Association of Plastic Manufacturers in Europe estimates that food wastage is as low as 2 per cent in 'developed' countries, but at levels of over 50 per cent in some developing countries.

The dilemma facing business managers is one of satisfying consumers' demands at the same time as accepting their responsibility to use more environmentally sound packaging. Due to consumers' changing lifestyles of the 1980s and 1990s, facilitated in part by rising consumer affluence, a demand was fuelled for convenience with quality; for example, single-serving disposable lunch packages, microwave dinners, fast food delicatessens and so on, all produce new pressures for packaging. Unless a dramatic change in consumers' passion for convenience and in the competitiveness of product displays in retail outlets take place, unnecessarily excessive packaging will continue to be used by marketers.

Case: Fast food: Our convenient, contemporary diet

Over the last ten years, many aspects of society have become increasingly dependent on so-called 'fast' (or 'junk') foods.

McDonald's, one of Tom Peters and Robert Waterman's (1982) 'excellent' companies (noting that managing our environment was not one of their criteria of excellence), became the target of environmentalists in the US in the late 1980s. While really no different from its competitors, as market leader, it was McDonald's polystyrene 'clamshell' burger box that became the illustration of some of the excesses of our disposable society. Lightweight polystyrene with its good insulating capability is a suitable packaging material, and a vital component for a fast food service. However, as highlighted in *The Economist* (29th August, 1992, page 62),

> . . . its use meant that a product which took seconds to consume was carried out of a shop in a package that would take centuries to rot.

With the US environmental pressure for greater responsibility likely to continue to grow, not only from lobbying groups, but also from employees and children (who are often the most enlightened environmentalists), reduction of their solid waste became an important corporate objective. Indeed, McDonald's signed a unique agreement with a major environmental lobbying group, the Environmental Defence Fund (EDF), to collaborate on solving this problem. If agreement for action was not forthcoming, it was decided that each organization would produce an independent report.

A range of alternatives were explored, including:

● recycle clamshells;
● replace clamshells.

McDonald's started with a recycling approach, under a 'McRecycle USA' programme. However, in practice, a number of operational difficulties became apparent, partly because the requirements for customers to sort their waste is time-consuming and partly because many customers take their food out of the branches to eat and therefore the waste is disposed of in a dispersed way. Moreover, a number of US cities decided to ban polystyrene packaging, so local franchises had to find alternative packaging to remain in business.

EDF believed that a strategy to reduce the amount of packaging was preferable to one of recycling. In the United States, the clamshell was replaced, where possible, by quilted paper packaging comprising a layer of tissue between a sheet of paper and a sheet of polystyrene. The replacement of clamshells, however, was not greeted by unanimous applause; perhaps not surprisingly, the polystyrene industry was unhappy. In a much-publicized claim, it was argued that the adverse environmental impacts of the relatively bulky clamshell are lower than the new wrapping made from virgin paper that could not be recycled. McDonald's commissioned its own research to compare the alternative forms of packaging, and it concluded in favour of minimizing waste at source, rather than attempting to collect and recycle it.

McDonald's corporate commitment to continue to reduce the adverse impacts of its business activities on the environment has received a positive response from customers and forced competitors to consider their attitudes and actions. The company has not stopped using the clamshell. It has a hierarchy of environmental goals as a means to achieve improved management of our environment:

● reduce;
● reuse;
● recycle.

The Economist's 'management brief' illustrates some of their actions:

● *reduce:*
  – packaging

  When it offered narrower drinking straws, customers found it hard to suck their milk shakes. But offering smaller napkins has not led to an increase in the number that customers use.

  – manufacturing impacts
    using less chlorine-bleached paper;
  – rubbish

  . . . by using plastics that are easier to recycle, or materials that can be composted.

● *reuse:*
  – transport packaging;
● *recycle:*
  – compost wasted food.

(Source: *The Economist*, 29th August, 1992.)

Under its controversial Waste Packaging Ordinance, the German Government is placing greater responsibility on both producers and retailers for product packaging, including their collection and recycling or disposal:

- from 1 December, 1991, they have had to accept all returned transport packaging;
- from 1 April, 1992, consumers have been able to abandon all packaging materials at the point of sale;
- from 1 January, 1993, retailers have had to take back all used packaging materials returned by consumers (and mandatory deposits have been levied on certain products to encourage the return of reusable packaging).

Manufacturers and distributors must take back their own packaging, or become a member of Duales System Deutschland (DSD), which is an organization established by German industry that grants licences to put the so-called Grüner Punkt (green dot) on packaging so that consumers know they can dispose of the packaging in DSD bins. Many foreign companies believe that these developments, sometimes called the 'Topfer Law' after Germany's Environmental Minister, place them at a distinct competitive disadvantage in the German market (see Figure 4.4 for Arthur D. Little's summary of the evolution of Germany's Green Dot programme). Moreover, at least in the short-term, the legislation has led to business collecting large quantities of waste without the domestic capacity to reprocess it (see also chapter five).

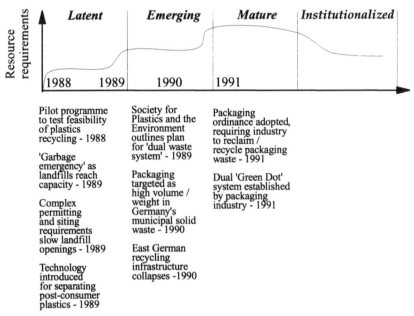

Figure 4.4 Evolution of Germany's Green Dot Programme
(Source: Blumenfeld, Earle and Annighofer, 1992, page 75.)

In 1994, there will be legislation in Belgium that penalizes organizations and industries which do not recycle their packaging products. Belgium industry believes it will cost them billions of francs. The levels of taxation will depend on what proportion of each product is recycled; for instance, the maximum tax of 15 BFr. will be imposed on a PVC bottle that cannot be recycled, while there will not be a tax on an 80 per cent recycled glass bottle. (See also the discussion in chapter five about the EC targets for the waste management of packaging.)

Case: Tesco: How green is green?

For all organizations, their impacts on the environment are not simply through their products and services, but the way they carry out their business.

Tesco is one of the largest and best known food retailers in the UK, with approximately 10 per cent of the nation's food market. There is no doubt that retailers are important 'gatekeepers' between suppliers and consumers, and, consequently, can play a significant environmental role. In countries where retailers hold a strong position, they have the bargaining power to demand high environmental standards from their suppliers down the chain.

For Tesco, the recent increase in consumer concerns for the environment provided an opportunity to develop the company's image by championing a new 'green' strategy. In 1989, for instance, Tesco came joint first (together with Safeway) in the Green Consumer's Supermarket Shopping Guide supermarket rating.

In early 1989, Tesco launched its 'Tesco Cares' campaign, with additional environmental information on its products, packaging, glass and paper waste recycling facilities. Furthermore, in part as an attempt to use less energy, the company has made some far-sighted changes to its distribution system by cutting the number of warehouses and reducing the movement of goods.

The business success and the publicity resulting from the above-mentioned changes in company policy, were followed by an advertising campaign under the slogan,

Tesco: The Greener Grocer.

However, Tesco's much-promoted product line of free-range chickens had its competitors crying foul because they were fed on an anti-parasitic drug!

Environmentally friendly products are offered as part of an extended product range, which still includes the environmentally unfriendly products to maintain consumer choice.

Price

Despite the increased importance of non-price factors in marketing, price remains a very critical element of the mix. As indicated in the previous chapter, an understanding of the full costs is necessary for their reflection in the final price of a product or service to the customer. To encourage more environmentally conscious consumption patterns, relative price is one of the mechanisms to stimulate modified consumer behaviour. Simply stated, it would be a straight forward matter if the prices of environmentally unfriendly products and services were higher than environmentally friendlier substitutes. For the environment, pricing can also be related to the issue of consumers' willingness to pay a premium price in their demand for environmentally friendlier products and services, such as organic products. In simple micro-economic terms:

How much of a premium price are consumers willing to pay for more environmentally friendly products (without a decline in product performance)?

On the other hand, because of lower taxes, the prices of environmentally friendlier products, such as unleaded petrol, are cheaper in order to stimulate changes in purchasing behaviour. Parenthetically, it is interesting to explore the extent to which unleaded petrol, which is more expensive to produce than an equivalent octane of leaded petrol, is becoming more popular because of being relatively cheap or because of being more environmentally friendly. In the past, when it was available at a higher price, there was little demand; to stimulate its consumption, Governments introduced differential taxes for leaded and unleaded petrol.

A 1990 Consumers' Association survey in Britain estimated that one in four people were put off buying organic produce because of higher prices; the current premium price is in the range 30–80 per cent, and for a successful market to be established this price gap will have to be reduced to 10–15 per cent. In the 1989 Mintel report on *The Green Consumer*, which was founded on a base of 933 adults, the anticipated price-demand relationship for organic foods is summarized in Figure 4.5.

More generally, as Table 4.3 demonstrates, in many countries, consumers say they are willing to pay higher prices to protect the environment. Respondents in a Gallup survey were told that 'increased efforts by business and industry to improve environmental quality might lead to higher prices for the things you buy', and then were asked,

Would you be willing to pay higher prices so that industry could better protect the environment, or not?

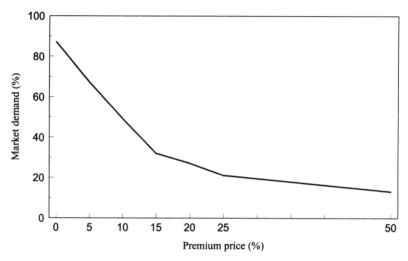

Figure 4.5 Anticipated Price-Demand Relationship for Organic Foods (Source: Mintel, 1989, *The Green Consumer*; sample size 933.)

However, there needs to be caution about the interpretation of such survey results. There is a significant difference between expressed attitudes and intended actions through changed behaviour. This situation is especially important in a recession time, when consumers have more short-term worries about getting value for their money, rather than contributing to saving the environment. Basic product features, such as quality, pricing and not least convenience, remain dominant drivers of consumers' purchasing behaviour. Indeed, more emancipatory market research, particularly concerned with the future, rather than the present or the past, is necessary to provide the required information. Marks and Spencer, for instance, completed an 18-month trial selling organically grown fruit and vegetables in 30 of their 280 stores, but decided to stop the experiment because sales did not provide a business justification. In part because of the recession, but also because of marketing hype, many environmentally friendly products have disappeared from the super-market shelves in recent months. Commentators are now arguing that 'green shopping' has passed its sell-by date!

The price premium is an important constraint. It is obviously too simplistic to see price as the only factor in a purchase decision, although its significance should not be under estimated. Interestingly, Mintel's *Second Green Consumer Report* in 1991 also considered non-buying behaviour. From a base of 1,336 respondents, the following three reasons were the main ones underlying a decision to not purchase:

- any environmental issue: 70 per cent;
- any ethical issue: 60 per cent;
- any animal issue: 60 per cent.

Table 4.3 Willing to Pay a Premium Price to Protect the Environment

Country	*Say they are willing to pay higher prices to protect the environment (%)*
India	56
Philippines	30
Turkey	44
Chile	64
Poland	49
Mexico	59
Brazil	53
Hungary	49
Uruguay	55
Russia	39
Republic of Korea	71
Ireland	60
Great Britain	70
Netherlands	65
Canada	61
West Germany	59
Denmark	78
United States	65
Finland	53
Norway	73
Japan	31
Switzerland	70

(Source: Gallup, 1992.)

Place

When marketers mention the third P, place, they refer primarily to branch location and distribution. A variety of marketing channels exist, with associated activities of physical distribution. Many public and private sector services are provided to customers through a network of branches over the country. The 'place' component of the marketing mix can be thought of as matching demand and supply interrelationships over space; there is a varied demand for different products and services from different households that can be satisfied by competing branches of various sizes or by direct marketing services.

Retailing, for instance, is said to be about 'location, location and location'. For managing our environment, it is especially relevant to discuss consumers' shopping trip behaviour and supply chain distribution channels for environmental management. There are important trade-offs between the numbers and sizes of branches and/or warehouses – a network of a few, central and relatively large branches/warehouses or a network of many, dispersed and relatively small branches/warehouses. The environmental impacts of such decisions are related not only to energy consumption, transportation costs and associated pollution, but also to local issues of congestion, noise and waste management. In the 1980s, for example, in many countries, traditional 'high streets' were being replaced by large out-of-town shopping centres. For an extensive discussion of consumers' shopping trip behaviour and associated branch location analysis, see, for example, Neil Wrigley's (1993) edited volume on *Store Choice, Store Location and Market Analysis*, and see the associated discussion in chapter five on distribution.

For both consumers' shopping trips and logistics, there is an increasing use and potential use of information and communications technologies as an effective enabling infrastructure. While the primary motive of such investments is not concerned usually with managing our environment, there can be direct benefits. For instance, the nature of retail banking, particularly the services provided at branch-level, is changing with the introduction of linked networks of automated teller machines and home banking systems; this trend is expected to continue as downward competitive pressures on costs are maintained.

Case: Teleshopping

For a few decades, the idea of shopping from home using information and communications technologies has been suggested as a potentially significant structural transformation for the retail industry. The success of Minitel in France is a much quoted example, and the home shopping shows in the United States seem to not only allow some customers to purchase a bargain, but also provide a new form of entertainment. Teleshopping is a term that covers a diverse range of means to provide remote shopping and service operations (see Ross

Davies and Jonathon Reynolds (1988) for more details and examples).

It is believed that home shopping will continue to develop as consumers demand more choice of both the products and services and their delivery. It is, however, misleading to view teleshopping as a replacement channel for traditional shopping behaviour; it will be a complementary form, and one that is used by only some consumer segments for some products and services. For instance, the initial, usually videotex-based teleshopping systems, were targeted at the time-poor, 'high' social classes, although the customers of home shopping shows are different with much greater focus on value for money. In terms of products, it is the straight forward ordering and delivering of essential goods that makes teleshopping attractive to some people.

The long-term success of teleshopping will be more dependent on product-related issues, such as product range, product quality, the speed of home delivery and the ease of making returns (if necessary), than on important system-related issues of ease of use, costs of use and reliability.

Promotion

The rise in environmentally conscious consumers has been accompanied by the rise in 'green' promotions. Advertising, public relations and personal selling make up the concept termed promotion, and it is the area of communications within the marketing function which has received most attention when related to environmental issues.

Traditional advertising is described by some commentators as fundamentally at odds with conservation and environmentalism, because the promotion of increased and excessive consumption inevitably has adverse affects on our environment. However, this perception is changing, as environmental pressure groups also see the power of the media for communicating their messages. For example, while the desire for environmentally friendly cars is causing significant product changes, performance is not really being compromized. Based on a 15-point questionnaire that was sent to thirty major car manufacturers, *Autocar & Motor*'s (1991) consideration of *Who's the greenest of them all?* derived for its British readers the 'top ten' of car manufacturers, which is summarized in Table 4.4.

Communications and information are the keywords. Clear and accurate environmental impacts must be communicated by organizations if the general public is to increase their awareness and make environmen-

Table 4.4 'Green' Top-Ten Car Manufacturers

- *VAG Group* Proving that a manufacturer can still be green even if it's big. Volkswagen and Audi remain the benchmarks when it comes to showing how the industry can clean up its act. The two other German companies ... areas advanced in the manufacturing and recycling processes, but neither has developed the green, mass market small car to the degree VAG has.
- *BMW* A worthy number two and close on the heels of VAG on the industrial side. But when it comes to the UK market, image triumphs over environment and advanced diesel versions of the new 3-series are still kept out. No small cars either.
- *Mercedes* Greener than most on the factory and recycling front, but it's hard to justify the new S-class. Commendable in all aspects apart from model range. Customers should expect greenness at this level.
- *Saab* Another front runner in the environmental stakes, even though resources are limited compared to the top three's. Industrially advanced and using the right materials, but serious recycling seems to be in the future. No diesels either. A forward-thinking philosophy keeps it just ahead of Volvo.
- *Volvo* Almost neck and neck with Saab. Perhaps even marginally ahead with recycling processes, thanks to volume of production. However, it loses a point to its compatriot because its greater resources should give it a clearer advantage.
- *PSA Group (Peugeot and Citroen)* – First non-German group to open a pilot recycling plant – definitely worth bonus points. Strong on electric power and diesel.
- *Ford* Ought to be doing more; if it is, it's not saying. Apparently it has a very 'green' philosophy on 'small cars' but does not seem particularly proactive, despite various 'concept' cars. Reformulated gasoline is an interesting concept – and perhaps Ford has the muscle to push oil companies and government for it. Will it, though?
- *Vauxhall* Not a lot new appears to be happening in the UK – but the Impact is a potential winner in the US. Ought to be leading the way, like it did with unleaded fuel in UK.
- *Renault* Still some way behind the Germans when it comes to recycling, but picking up points for its innovations in that area with the new Espace, and for its work on electric power fuels and engine research.
- *Fiat* Another mega-group which should be higher placed. But green issues – far more emotive to its exports market than its domestic one – are being taken on board in Turin. Strong on small cars and alternative power sources, getting better with the manufacturing process. Recycling is still only a glimmer on the horizon.

tally sound purchasing decisions. Notwithstanding the issue of misleading promotions, some people argue that the majority of current environmental promotions are aimed at specific, targeted niches, rather than the general population or mass market. Moreover, the promotion of a 'green' product seems likely to be less successful in a number of different situations:

- for luxury products;
- for products that do not satisfy traditional performance standards;
- for products for which a strong and meaningful brand image cannot be developed.

Unfortunately, although the advertising industry has voluntary codes of practice, misleading advertisements still seem to find their way into marketing campaigns. In Britain, recently, a number of promotions have been attacked for deceptive, inappropriate or unsubstantiated claims. The Friends of the Earth, for example, have established an annual 'Green Con' award:

- British Nuclear Fuels won the first 1989 'Green Con' award for its advertisement 'greenhouse effect';
- Eastern Electricity, a recently privatized regional electricity company, won the 1990 'Green Con' award for urging the public to ... *use more electricity rather than less* as a means of combating global warming, because ... *there is very little carbon dioxide or global warming gas emissions from the all electrical home*;
- Fisons won the 1991 'Green Con' award for claiming that peat extraction did not endanger Britain's remaining wetlands of conservation value, even though 90 per cent of their peat-cutting operations are on peatbogs officially designated as Sites Of Special Scientific Interest (SSSIS);
- Meyer International, owners of the Jewson builder's merchant chain, won the 1992 'Green Con' award for misleading and inaccurate statements about the role of the timber trade in the battle to save the world's forests, especially the tropical rainforests.

The UK Independent Broadcasting Authority recently banned a £3.3 million advertising campaign for Scott's Andrex toilet rolls in which they claimed their reforestation,

... helps to combat the greenhouse effect.

As the above examples indicate, some advertising is still confusing and potentially deceiving, and there is a need for more accuracy and responsibility. A 1989 Consumers' Association survey, which appeared in their January, 1990 *Which?* magazine, found British shoppers confused about product labelling, with over half the respondents believing that some form of official approval had been granted already. The report also describes different types of erroneous claims:

- excessive claims;
- multiple claims;
- incomprehensible claims;
- meaningless claims.

Until general product 'ecolabelling' exists to specified standards, the most environmentally friendly products may still lose out to competitors. Eco-labelling, which was first introduced in West Germany in 1977 with the 'Blue Angel' system covering about 3,500 products, is concerned about environmental acceptability within a particular product category. It is an illustration of a market-oriented environmental policy by attempting to influence consumer behaviour through an identification of the relative environmental impacts of different products. Notwithstanding the real progress made by different schemes, it must be appreciated that they are not fully comprehensive because the label is often awarded on the basis of a focus on particular environmental characteristics.

Eco-labelling schemes now exist in other countries, including Norway, Sweden, the United States, Canada and Japan. While the majority of the schemes are administered in the public sector, in the United States, the two schemes, 'Green Cross' and 'Green Seal', are both administered by private sector organizations.

In Britain, environmental pressure groups, such as the Friends of the Earth, have indicated that they would like to see an environmental labelling scheme controlled by the Government and with a long-term, 'lifecycle' analysis of how individual products affect the environment. Such a scheme has been recommended by the House of Commons Select Committee on the Environment, but the Government is waiting for further European Community legislation. In 1991, the EC proposed to introduce a pan-European eco-labelling system, named "Eco-audit", which would be based on the life-cycle concept. This planned EC ecolabelling system will initially be voluntary, but it should cause many managers to re-examine their business strategies. It is likely to exclude food, drink (to avoid consumers' confusion between environmental impact and nutrition) and pharmaceuticals. Table 4.5 provides a list of the product groups for which criteria are being developed.

The scheme has two basic objectives:

- to encourage manufacturers to design and produce products which have a reduced environmental impact during manufacture, distribution, consumption, use and disposal;
- to provide consumers with better information on the environmental performance of products.

Table 4.5   EC Eco-labelling Scheme

Product group	Lead country
Paper products	Denmark
Textiles	Denmark
Insulation materials	Denmark
Laundry detergents	Germany
Paints and varnishes	France
Packaging materials	Italy
Washing machines	UK
Dishwashing machines	UK
Light bulbs	UK
Soil improvers	UK
Hairsprays	UK
Growing media	UK
Hair styling aids, antiperspirants, deodrants	UK
Dishwasher detergents	Germany
Household cleaning products	Germany
Batteries	France
Shampoos	France
Refrigerators	Italy
Floor and wall tiles	Italy
Cat litter	Netherlands

Member states have to designate a national point of access to the scheme. A UK Eco-labelling Board (UKEB), a non-departmental public body, has been established with responsibility for:

● proposing product groups for inclusion in the scheme;
● assisting in the development of environmental criteria for products;
● assessing individual applications for a label;
● concluding contracts with successful applicants;
● promoting the scheme and handling enquiries from the public and business.

It is only with full information that consumers can be expected to understand the environmental consequences of their purchasing decisions. Eco-labelling is likely to give a commercial boost to truly green products and services. It is, however, only a first stage in improving consumers' environmental knowledge of different products and services. The publication of independent consumer guides is also helpful, and the significance of retailers as 'gatekeepers' should not be underestimated. Retailers can influence not only their customers, but also their suppliers.

(Data) Processing

> Good information is a facilitator of successful marketing, and indeed, seen in this light, marketing management becomes first and foremost an information processing activity.
>
> (Christopher, McDonald and Wills, 1980.)

The issue of data as a corporate resource and the need for more and better information for decision-making are considered more fully in chapter six on information resources management. In the context of marketing, it is important not only to appreciate communications about products and services to potential consumers through promotion, but also to understand the requirements to effectively handle information about customers, products/services, markets and competitors.

The nature of decisions affects the type of information needed, and, therefore, also the design and specification of a computer-based information system. The solution is not technological; it is dependent on management's awareness of the real potential. Recent research, for example, by Oasis (1989), indicates that the handling of marketing information is relatively poorly developed. Indeed, it is far from clear that information systems have necessarily been useful and relevant for making marketing decisions. Table 4.6 is illustrative of some of the information requirements for some marketing tasks.

(Marketing) information systems must support not only management's decision-making, but also relationships with customers, suppliers and possibly collaborating competitors. As the move towards 'maxi-marketing' continues, the strategic and tactical significance of customer databases will increase. For managing our environment, this should enable more targeted promotion of particular products and services, as well as effective educational promotions.

Marketing revisited

The argument introduced at the beginning of this chapter, that radical changes in patterns of consumption are essential if we want a sustainable future, is bound to have direct consequences for the very essence of the marketing function in the future. The underlying aim of environmentally conscious consumerism should be to reform, rather than restructure, patterns of consumption. As a consequence, the nature of strategic marketing planning in the 1990s will be much different from the 'simpler' business environments of the 1960s, 1970s and 1980s.

Table 4.6 Information and Marketing Tasks

Task	Information requirements

Analysing the market	Sales/profit:
	    total
	    by products, area, client
	    market rates of growth
	    cash flows by each segment
	Client attitudes:
	    brand awareness
	    brand loyalty
	Number of customers:
	    by product
	    by area
	    by purchasing patterns
	Units sold per unit input:
	    advertising to sales ratio
	    personal selling effort to sales
	    shelf space to sales

Defining marketing objectives	Sales/profit performance Market share Cash flows Competitive strengths/weaknesses Technical, legal, political and social influences on the market Resources and skills in the organization Buyer loyalty Channel loyalty Customer needs and buying power Financial position Manufacturing competencies, capacity and flexibility Research and development strengths

Developing appropriate marketing strategies	Data on current strategies:
	    cost
	    effectiveness
	Identification of strategic options:
	    cost
	    potential effectiveness
	Targets
	Budgets

Controlling performance	Units sold per unit input:
	    advertising to sales ratios
	    personal selling effort to sales ratios
	    shelf space to sales ratios

(Source : Parkinson and Parkinson, 1987.)

Marketing is an increasingly important tactical and strategic function of all organizations. For progress on managing our environment, it is essential to re-examine many of the conventional wisdoms of marketing. Indeed, the argument extends like accounting and finance into the area of 'corporate governance', especially in connection with business ethics and corporate responsibility (see also the discussion in chapter seven).

The discussion in this chapter has centred around the impacts of increased environmental awareness on the marketing mix, and a major challenge facing business seems to be the question relating to ethics in marketing. The step of moving from using marketing as a promotional exercise, particularly for green, cosmetic purposes, to real environmental concern is vital. It is becoming increasingly insufficient just to create an environmental image. Credibility is important, and it must also be sustained and believed by customers in order for business to be successful in the long run.

In short, environmentalism must be perceived as a business cause and not as a marketing scheme. The Body Shop is an excellent illustration of how such a policy can be carried out in practice. The Body Shop has been cited as 'a paradigm for how to sell in the Nineties', a philosophy towards selling, with a tell all tradition which is so contradictory to the way companies are used to view marketing. Why does this strategy work? According to *Fortune* (13 January, 1992), it is because,

> . . .typical Body Shoppers are at the back of the baby boom, a sceptical group. They distrust advertising and sales hype, demand more product information than their elders, and are loyal to companies they consider responsible corporate citizens. Marketers believe they represent the future and that their business will go to companies like Body Shop that develop a forthright relationship with them.

Case: Our environment: Consumers' sovereignty

As people become more informed about the environmental implications of their consumption patterns, through their consumer sovereignty, real opportunities for managing our environment should exist.

Craig Smith's (1990, pages 293–295) research on ethical purchase behaviour offers a number of interesting and relevant conclusions. His five key findings are:

1. *The recognition of ethical purchase behaviour.*
   Consumers may be influenced in purchase by what may be broadly described as ethical concerns. This behaviour seeks social responsibility in business as defined by the consumer.

2. *Consumer boycotts should be judged as symbolic acts as well as on their effectiveness; the former may be more important in their success.*

There is a distinction between boycott effectiveness and success; the two do not necessarily go hand in hand. Boycotts also have expressive functions as well as instrumental functions. Accordingly, there are two types of boycott: the symbolic and the effective. The former achieves compliance with demands made through moral pressure, the latter through economic pressure. Factors in boycott effectiveness and success can be identified according to the choice of target, pressure group organization and strategy, and response to the boycott, but conditional factors vary. It is only possible to refer to necessary and sufficient conditions for a successful consumer boycott at a generalized level. They are: support on the issue and consumer sovereignty. Put otherwise, consumers must be concerned, willing, and able to act. However, boycotts should not be judged in isolation from other tactics likely to be employed by pressure groups alongside the boycott.

3. *Management response strategies to consumer boycotts are: ignore, fight, fudge/explain, comply; although a proactive strategy, in anticipation of increasing pressure group activity, has most to recommend it.*

The proactive strategy involves management co-operating with pressure groups. Within an identification of those market segments containing consumers concerned about particular ethical issues, it attempts to enhance the legitimacy element in the firm's marketing mix and thereby avoid negative product augmentation.

4. *There is a role for pressure groups in the marketing system in the social control of business.*

Pressure groups act as a countervailing power to business. In the marketing system they enhance consumer sovereignty by providing the necessary information for ethical purchase behaviour. They may organize ethical purchase behaviour in the form of consumer boycotts. There are three forms of social control of business according to the type of power involved: legislation (force), market forces (inducement), and moral obligation (manipulation). Pressure groups, in using the consumer boycott, are attempting to employ the market form of control. In symbolic boycotts, however, control is more a result of moral obligation working through the market.

5. *The domain of consumer sovereignty is only limited by information and choice.*

Ethical purchase behaviour suggests that consumer sovereignty may have two dimensions: degree and domain. Domain refers to the jurisdiction of consumer sovereignty. It depends on choice through competition and, especially, information. The importance of pressure groups in the marketing system stems from their provision of information on social issues. Where consumers act on this information in purchase then there is political participation or purchase votes in the market-place. Yet the possibility or threat of ethical purchase behaviour, acting as a constraint on firms, is important in itself for the social control of business.

Concluding comments

Some of the success of marketing can be associated with the general, increased emphasis on material values. This outlook can create detrimental pressures on our environment. Recently, there have been some attempts to market environmentally friendlier products, as well as even to encourage less consumption. While there is growing empirical evidence of some changing consumer behaviour for the environment, the scale and permanency of any shifts remain unknown and can still be questioned. Complacency could be a real danger. Consumer sovereignty and corporate social responsibility must be inextricably linked, if marketing guru Philip Kotler's (1980, page 35) 'societal marketing' concept is to have real meaning in practice,

> ... a managerial orientation that holds that the key task of the organization is to determine the needs and wants of target markets and to adapt the organization to delivering the desired satisfactions more effectively and efficiently than its competitors in a way that preserves or enhances the consumers' and society's well being.

In summary, environmentally conscious marketing must incorporate broader and strategic relationships, both internal and external, rather than tactical-level actions and communications. Perceived, short-term business benefits may occur from cosmetic environmental actions, but competitive advantage will only accrue if there is a set of committed and responsible policies and actions that go beyond individual products and services to cover all an organization's activities.

It would, however, be misleading to assume that greater environmental consciousness will undermine the traditional role of the marketing function. Indeed, the promotion of customer service and product quality, including efficiency, longevity and reliability, should be fundamental objectives of marketing.

In terms of this consideration of marketing, it is relevant here to question the suppliers', rather than the consumers', perspective that is given by the traditional Ps in the marketing mix. Individual consumers are affected by the characteristics of any product or service. Following Theodore Levitt's 'augmented product concept', people do not simply purchase products and services, they purchase the expectation of benefits and costs. A cost, for example, may be a legitimacy cost for consumers with consciences for our environment.

By reflecting on the traditional concepts and principles of marketing, it is clear that consumer sovereignty is a fundamental underlying force, and that environmental concerns will become an ever more significant influence behind consumers' purchasing behaviour. Consumer sovereignty, depending on its extent in practice, provides a clear opportunity if there is real concern for our environment; however, it should not be equated merely with consumerism. In one sense, every purchase or non-purchase can be interpreted as a vote for or against the environment. Michael Baker (1974, pages 281–282), for instance, highlights that,

> ... the nature of marketing and consumerism reflect a fundamental paradox for while they are invariably seen as being in conflict both activities possess the same objective - customer satisfaction.

Craig Smith (1990) indicates possible 'consumer pressure for corporate accountability' in his wide-ranging examination of *Morality and the Market*. However, unless there is complete consumer sovereignty (and good information), the market cannot provide all the choices and controls. The reverse of the business opportunities argument is that the markets for environmentally unfriendly products will decrease or disappear, as exemplified by the reductions in the demands for chlorofluorocarbons and for phosphates in detergents.

Review questions

1. Describe the traditional marketing mix, and indicate, through examples, ways in which it needs to be modified for a more effective management of our environment.
2. For a selected product, examine in detail its current packaging, and consider alternative options that would be better for the environment.
3. What roles do, and can, retailers play for enhanced management of our environment?

Study questions

1. In a country of your choice, explore the development of its 'green movement' over the last quarter of a century.
2. Select a specific product or service, and consider the extent to which competing companies use environmental consciousness as part of their promotion strategy.
3. Explore the national and international dimensions of eco-labelling.

Further reading

Charter, Martin (ed) (1992) *Greener Marketing*, Greenleaf Publications, Sheffield.

An edited volume of a consistently high standard, which provides a clear discussion with a cogent argument for a more responsible approach to business. The chapters are organized around the strategic and practical implications of greener marketing (although greener marketing is interpreted in a broad management, rather than specific functional, sense); these provide the firm foundation for twenty short case histories:

- *Strategic Issues:*
    - pressure group perspective: Friends of the Earth;
    - evaluating green performance: Merlin Research Unit;
- *Corporate Response:*
    - opportunity from threat: ICI;
    - greener retail and supplier audits: B&Q;
    - a Canadian retailer's perspective: Loblaw Companies Limited;
    - environmental policy development and implementation: Exel Logistics;
- *Greener Marketing Strategy:*
    - open communications: British Gas and McDonald's;
    - balancing ecology and economics: Countryside Holdings;
    - trading for a fairer world: Traidcraft;
- *Greener Products:*
    - a nick in re-manufacturing: Onyx Associates Limited;
    - supporting sustainable development: Issan Mori;
    - developing greener substitutes: Dawes Environmental Coatings;
    - the importance of research: Natural Fact;
- *Greener Packaging:*
- - greener packaging policy: Gateway;
- *Greener Logistics:*
    - integrating environmental policy: The Lane Group;
- *Greener Pricing:*
    - socially responsible profits: The Ethical Investors Group;
- *Greener Communications:*
    - involving the stakeholders: Hampshire County Council and Sutton Borough Council;
    - eco-sponsorship: The Green Business Service;
- *Greener People:*
    - promoting organizational change: British Telecom;
    - the importance of internal marketing: Pilkington Glass Limited.

Smith, N. Craig (1990) *Morality and the Market*, Routledge, London.
While a difficult read, with many indications of being a modified PhD thesis, Craig Smith provides a sound theoretical and practical overview

of ethical (ecological, ideological and political) consumer behaviour, particularly with regard to the key element of consumer sovereignty and the specific issue of consumer boycotts. Smith argues that 'consumer pressure for corporate accountability' is consistent with the traditional marketing perspective of customer satisfaction. Ecologically concerned consumption is discussed briefly as an illustration of ethical purchase behaviour, and, in the preface (page ix), the author notes that,

> Ethical purchase behaviour even extends to the Royal Family. Prince Charles received considerable press coverage for his comments earlier this year that aerosols are banned from the Royal household. He is concerned about the damage to the ozone layer caused by the chlorofluorocarbons (CFCs) in aerosols.

Stilwell, E. Joseph, Canty, R. Claire, Kopf, Peter W. and Montrone, Anthony M. (1991) *Packaging for the Environment*, Arthur D. Little Incorporated, New York.

> Packaging is the ultimate symbol of our consumer culture. It tells the story of our technological achievements, preserves our food, protects what we buy, and raises our standard of living. It plays a vital and growing role in the global economy. And through Andy Warhol's vision, the Campbell's soup can and the Brillo box have even been elevated to an art form. At the same time, packaging is also the largest single contributor to one of the nation's most troubling environmental problems: the municipal solid waste crisis.

A very detailed and clear examination of the need for attention to the scale and nature of contemporary packaging, because the planet's capacity to act as a disposal site is getting much smaller. To create an industrial ecosystem, progress through partnership is essential. It is a clear message for business, with direct business benefits envisaged from responsible actions. The arguments are supported by up-to-date examples, and the book ends with five case studies:

- pollution prevention: 3M Company's 3P program;
- a leadership role: Du Pont's environmental policy;
- choices for consumers: the Proctor & Gamble approach;
- Johnson & Johnson: world-wide environmental responsiveness;
- a venture in environmentalism: the natural polystyrene recycling company.

5 Production and operations management

Corporations that achieve ever more efficiency while preventing pollution through good housekeeping, materials substitution, cleaner technologies, and cleaner products and that strive for more efficient use and recovery of resources can be called 'eco-efficient'.

(Schmidheiny, 1992, page xii.)

Introduction

Many of the visual concerns about the adverse impacts of business activities on our environment relate to industrial production processes, the pollution and associated waste. In defining the function of production and operations management, Keith Lockyer, Alan Muhlemann and John Oakland (1991) highlight five, interrelated Ps to be managed:

- Product;
- Plant;
- Processes;
- Programmes;
- People.

Figure 5.1 summarizes the interrelationships of these five Ps.

Within the context of an organization, the production and operations management function only exists to satisfy customers, with their needs flowing into an organization, which, in turn, transforms them into a product/service that satisfies them. The interdependencies between the marketing and production and operations management functions, particularly with regard to the products' specification and delivery, are vital to an organization's success. Market-driven demands, understood as the customers' wants and needs, must be communicated to the people

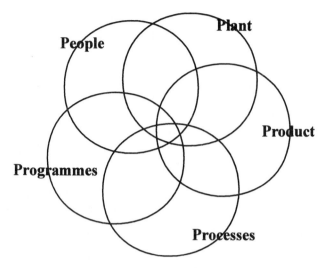

Figure 5.1 The Production and Operations Management Function

responsible for the production process. As a simplified representation, Lockyer and his colleagues idealize the process as five, sequential input-output diagrams (see Figure 5.2). The function of production and operations management comprises the 'analysis', 'supply' and 'transformation' facilities.

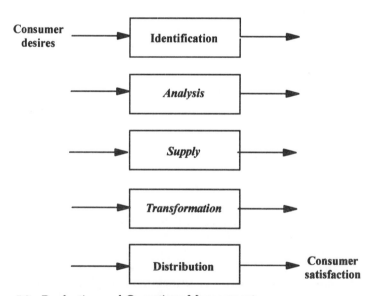

Figure 5.2 Production and Operations Management

The output, or the 'product', of this transformation process can be either tangible or intangible. Management of this functional area is dedicated to the organization of the use of facilities, equipment and other resources.

Within production and operations management, perhaps as to be expected, conventional practice focuses on the utilization of the cheapest materials and processes – maximum output with minimum input of resources. Moreover, large-scale production has been justified in terms of economies of scale. However, measurement or inclusion of the problems or the costs of environmental degradation and depletion commonly caused by such production activities are incorporated infrequently.

With increasing, and very often changing, legal and public environmental pressures, the requirements for production and operations management are being changed. The production and operations management function must now respond to new and more environmentally friendly market-driven demands, in terms of both:

● the products;
● the associated production processes.

From an environmental perspective, the use and consumption of many products, such as cars, hardware furniture, and so on, are not environmentally friendly. Moreover, the processes by which they are produced are often not environmentally friendly. For example, the creation of pollution, water, air and noise pollution, and waste are clear and serious side effects of many production processes. The societal loss to both the environment and to people resulting from a manufactured product can include:

● consumption of resources, especially non-renewable or non-substitutable, in the production process;
● the costs of any environmental damage resulting from a product's use;
● the detrimental effects on people's health from production processes, product use and associated pollution.

Stages of environmental action for production and operations management

In recent years, the narrow orientation of concerns has shifted from the depletion of natural resources to include the adverse polluting impacts of industrial production processes, product use and product waste. A number of stages of environmental action for production and operations management can be differentiated, including:

● end-of-pipe solutions;
● clean technologies;
● cradle-to-grave management.

Today, pollution far exceeds the carrying capacity of the planet, but it would be unrealistic to believe that the world can do without manufacturing. Moreover, particularly in today's economic recession, many organizations, including large and well-known organizations, cannot afford to make the investments in new facilities and equipment to become more environmentally friendly. For instance, the Business Council for Sustainable Development argues that there are three significant obstacles to a broader adoption of pollution prevention by industry:

● economic constraints;
● lack of information;
● management attitudes.

Thus, it is important to discuss how and why changes towards environmental excellence in traditional production and operations management are necessary, and also to point out some of the changes that are already taking place. There is a need to change current perspectives from a so-called 'end-of-pipe' solutions by taking a broader and integrated approach to production and operations management that explicitly considers the environmental implications through a product's entire life cycle from cradle-to-grave.

In some sense, it is necessary for new, cleaner technologies to also lower production costs, and this fact must be demonstrated clearly. Progress can usually be made initially by enhanced housekeeping and by making the existing processes more efficient. Eventually, however, there probably needs to be some technological change, and, therefore, significant investment. Lack of comprehensible information about the various alternatives is also an important constraint, and the availability of 'demonstrator' or 'lead' examples would be especially useful. Ultimately, managers' attitudes need to be more positive, appreciating that the long-term business benefits can outweigh the costs. Their real concerns focus usually on short-term costs, which become the basic driving force behind the adoption of 'end-of-pipe' solutions, rather than the prevention of pollution generation.

Traditionally, efforts to curb pollution in production have focused on end-of-pipe solutions, removing harmful pollutants in the waste streams coming out of a plant after they have been created. This policy does not imply any change in the traditional production (or business) processes. Waste is not eliminated, and pollution is transformed basically from one type to another type that is currently considered more safe and less harmful. This type of pollution control can be viewed as a tactical operational expediency, which does not address the problems at a longer term, strategic level. Simply stated, environmental management is not incorporated into the organization's overall corporate strategy. However, end-of-pipe solutions remain the most commonly used, because 'add-on' technology is the cheapest, at least in the short-term. It is also noted that investments in end-of-pipe equipment are often undertaken merely to comply with legislation.

The next stage is towards the use of clean technologies, which require fewer natural resources and create less pollution as well as less waste. This is a move from clean up ('reaction and repair') to pollution prevention and waste minimization at source. A clean technology could for instance include greater use of non-toxic production methods. This approach has become worthwhile implementing for some types of organizations, often due to increasingly more stringent legislation and public pressures. Moreover, as the costs of polluting and producing waste, as well as the benefits from being clean, continue to rise, financial benefits can accrue to organizations from such investments.

Today, some organizations are recognizing that their responsibility does not stop with the production or sale of a product. A 'cradle-to-grave' stewardship perspective is being adopted over a product's entire life cycle. This indicates the adoption of a comprehensive ecological view of a product's impacts on the environment, commencing with R&D and design through the extraction and use of raw materials, production and processing, storage, distribution and use, to the final disposal of the product and the waste created as a by-product. This view implies a full and integrated consideration of all the environmental impacts.

Figure 5.3 summarizes a PA Consulting lifecycle analysis of the environmental impacts of washing machines across the four main phases of production, distribution, use, and disposal. Such life cycle principles demand a radical change in the prevailing production and operations management perspective. From a relatively adverse position at the beginning of the 1980s, AEG's 'green' washing machine and the organization's commitment to quality created a successful business repositioning in the top price niche of the market (see Figure 5.4).

In this chapter, our discussion is centred around the cradle-to-grave management philosophy, investigating each of the components in the life cycle, and it is structured according to Figure 5.5. The focus is on manufacturing, that is the production and operations management processes of making physical goods. Operations management of services should be seen as similarly important, and the life cycle philosophy is

Percentage

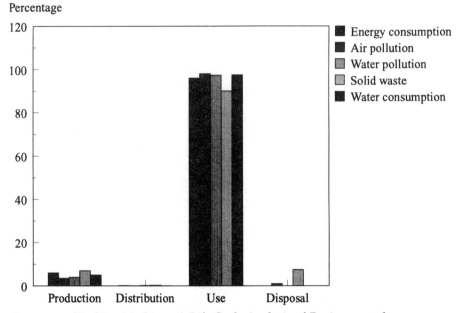

Figure 5.3   Washing Machines: A Life-Cycle Analysis of Environmental Impacts

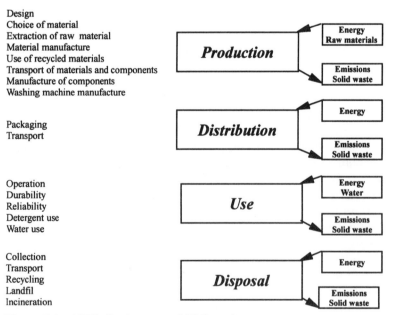

Figure 5.4   AEG's Environmental Philosophy
(Source: adapted from Charter, 1992, page 173.)

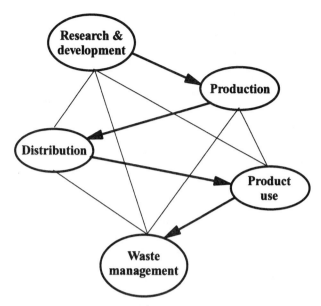

Figure 5.5 Cradle-to Grave Management

certainly applicable to the management of an intangible outputs. For example, the location of retail branches and their provision of different products and services have direct and indirect environmental impacts.

While, for convenience of presentation, it is easier to discuss each component in turn, it is stressed that the interrelationships between all of them must be understood as the underlying principle of cradle-to-grave management. Changes in one area will inevitably influence the others. For instance, a higher quality, albeit apparently higher priced, input raw material may have a beneficial impact on the maintenance and repairs required during the product's useful life, thereby reducing the overall resources needed. Although the significance of each component in Figure 5.5 is discussed, in practice, to date, most attention has been devoted to production processes and waste generation and disposal. First, however, there is a brief discussion of total quality management, specifically its consistency with managing our environment (see also the discussion of BS7750 in chapter six).

Total quality management

Environmental management can, and should, be linked to the concept of total quality management (TQM), which was developed in the 1980s. Attempts to improve an organization's environmental performance should be seen as part of either a cost-based or a differentiation-based business strategy. For example, since its establishment in 1975, it is estimated that 3M's well-known Pollution Prevention Pays programme is now saving $500 million per annum; Epson had a total quality programme called *Do we really have to use CFCs?* Indeed, the Japanese quality guru, Genichi Taguchi, emphasizes the need to minimize this 'societal loss' from production (see also Ledgerwood and colleagues' (1992) argument for a total quality approach to link the environmental audit and business strategy).

Working with the United Nations Environmental Programme, the Global Environmental Management Initiative (GEMI) promotes environmental excellence to the global business community through two stages of implementation:

- by enhancing awareness of businesses' responsibility;
- providing information and assistance to companies that want to enhance their environmental performance (linked to business performance).

TQM ultimately aims for zero defects, preventing defects occurring in the first place, not only in the product or service, but also at every stage of the production process, both internally and externally. This is a responsibility for quality that is shared by everyone.

Significant parallels can be drawn between aiming for total quality and the concept of cradle-to-grave environmental management. The environmental equivalent of zero defects in TQM is the ultimate goal of zero adverse impacts on the environment. Some organizations now believe that the only completely safe environmental option is to remove pollution completely, noting, for example, Du Pont and ICI's decisions to stop the production of CFCs. However,

Are zero environmental impacts realistic and therefore achievable?

As zero defects in TQM might be unachievable in many industries, zero harmful environmental impacts are probably unachievable even in a sustainable society. On the other hand, while emissions and waste are inevitable consequences of many human activities, it should not mean they have to be environmentally harmful. Instead, it could be argued that emissions and waste should be confined to the forms and amounts the local environment can accommodate, the so-called threshold-effect. Thus, environmental impacts can, and should, be reduced, but cannot realistically be eliminated completely. The objective of zero impacts on the environment, defined broadly, can never be met by most organizations, although continuous improvements in the TQM philosophy are realistic.

Cradle-to-grave management

In this main section, the discussion is structured by the following components that are linked inextricably in production and operations management:

- Research and Development;
- production;
- distribution;
- product use;
- waste management.

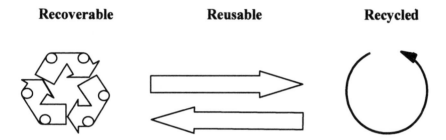

Recoverable　　　　　**Reusable**　　　　　**Recycled**

Figure 5.6　Liquid Waste Disposal

At the end of this main section, a long case about the motor car is used to illustrate the philosophy and scope of cradle-to-grave management. In chapter six, there is a more extensive discussion of the approaches to life cycle analysis.

Research and development

A fundamental area where improvements in environmental management are necessary is in the design and development of products, as well as the specification of plants and their processes. Research and Development (R&D) is a vital activity for the identification of new and environmentally friendlier materials, products and process technologies. The design of products should be closely linked to the impact on the environment from their production, use and disposal, with attention on aspects such as increased energy efficiency, quality, durability, recyclability and the use of less harmful materials.

A well-known example of R&D for product replacement is the use of CFCs, chloroflourocarbons, which is one of the major contributors to the ozone layer depletion. CFCs are used in approximately five thousand different applications to produce goods and services worth about $30 billion a year. More than $200 billion worth of already installed equipment worldwide depends on CFCs. CFCs are mainly used for refrigeration, foam plastics and aerosols. Aerosol manufacturers have already phased out CFCs in for instance deodorants and hair sprays. The electronics industry has used CFCs in cleaning fluids, but it has to a large extent now substituted the CFCs with water-based cleaning fluids. However, the most important application of CFCs (about 40 per cent of all CFC consumption) is in all types of refrigeration: domestic fridges and freezers; air conditioners; supermarket chill cabinets; and so on. The chemical industry is now developing a new 'family' of fluorocarbon compounds, HCFCs and HFCs, to replace CFC refrigerants. On the other hand, the new flourochemical replacements require more complicated production processes. Additionally, the manufacturers have substantial sunk investment costs to recoup. As a consequence, the substitutes are more expensive than the CFCs they replace. The market is still too new to give a clear indication of prices, but it is likely that the substitutes will cost three to five times as much as CFCs.

An example of an organization pursuing a company goal of zero pollution in all its activities is the chemical giant Du Pont. One illustration is the company's voluntary suspension of all production of CFCs by the millennium, a $750 million a year business for Du Pont in which it is the largest producer. Expensive R&D for safe compounds to replace CFCs in cleaning, refrigeration and other uses is being undertaken. To date, Du Pont states it has already spent over $240 million in developing CFC alternatives, and the company estimates that expenditures could exceed $1 billion over the next decade. Interestingly, increased environmental sensitivity is evidenced everywhere in the organization, partly because this dimension has been introduced as one of the criteria determining managers' compensation.

Design relates not only to the core product, but also to what in popular marketing terms is named the tangible product, which includes packaging. As mentioned in chapter four on Marketing, packaging often constitutes a major part of the total product and its environmental impact. Reduction in packaging is the ultimate goal, with sound packaging principles, such as lightweighting, recyclying, alternative material selection and others, being important alternatives.

For environmental reasons, design must also involve the process side, a design issue which unfortunately is very often forgotten through a narrow focus on product design. That is, there is a design dimension in the development and construction of plant and equipment. Environmental effects of production must be estimated and subsequently evaluated and monitored, including matters such as optimal plant efficiency, safety, and clean and quiet operations.

More specifically, plant location is becoming increasingly important with rising public pressure for preservation of greenfield sites and the NIMBY ('not–in–my–back–yard') syndrome. The EC Environmental Impact Assessment directive is likely to be increasingly significant, by which management takes a view of the impacts of, for instance, production facilities on the environment, prior to the commencement of a project.

From a competitive perspective, the issue of 'time-to-market' can be a vital competitive source, and there are no reasons to doubt that organizations that recognize this fact in environmental R&D and innovation should accrue business benefits. Flexible manufacturing and rapid-response systems are important requirements for customer service.

Case: Is the chemical industry attempting to clean up?

In a recent survey of some of the largest multinational chemical companies, the lack of planned, environmentally-related R&D is striking. Most companies do not give routine attention to environmental aspects in their R&D planning, and only some organizations have environmental projects on an ad hoc basis. Additionally, there appears to be no systematic selection of new products because of environmental objectives.

It is interesting to note that, in the development of new products, the environmental strategies adopted by these companies are far from radical:

● The majority of work of R&D departments (over 80 per cent) is paid for by existing product groups leaving only a small percentage of funds for independent research.
● One product strategy is a 'backing two horses' strategy. This relates to an approach where both a short-term and long-term solution for a specific environmental problem is sought. The short-term is a relatively easy modification of the existing product aiming at bringing in as much money as possible from existing products. At the same time, a more fundamental product modification to resolve the problems is developed. However, in order to avoid cannibalization of an organization's existing market, the long-term alternative products are often consciously kept at a laboratory stage.
● Another, related strategy variation is a 'diversification strategy'. This relates to a strategy where a more ecologically sound product is

developed but is priced so high that it cannot compete directly against the cheaper, existing (and more environmentally unfriendly) product. In this way, the companies are able to create two markets for their products, hoping to increase their total market share.

Due to this situation regarding the development of new products, where the chemical companies are not pressing to substitute existing products, most of the companies believe that the existing production patterns will continue as they are for ten to twenty years. At the same time, however, it is realized that,

> . . .aspects of the environment would start to play a more dominant role in the structure and culture of R&D in the near future.

It should be mentioned, however, that the chemical industry is experiencing big increases in spending, relating to other areas of production and operations management, in order to improve pollution control. Efforts are made world-wide by the chemical companies to clean up what in retrospective is acknowledged as being their appallingly negligent treatment of the environment.

All international chemical giants have increased their expenditure on pollution control over the past decade, a fact which is particularly true for the German, Dutch and Swiss companies which face especially tough legislation. ICI, the leading UK chemical group, announced in November 1990 that, until 1995, it would double its environmental expenditure to £1 billion worldwide. In 1985 ICI's environmental spending was £50 million, in 1990 £125 million, and, for the period, 1991 to 1995, an increase to an annual spending of £200 million is planned. This figure will account for more than 20 per cent of the group's total capital expenditure. As with many multi-nationals, this environmental commitment is global, with no double standards of operation in different countries, particularly developing countries. Moreover, this strategic decision was taken in ICI during a period of general economic recession with declining profits (and a possible hostile take-over from Lord Hanson).

Environmental protection remains focused on 'end-of-line' solutions. However, during the 1990s, the emphasis will switch gradually towards altering the manufacturing processes themselves, that is towards a 'cradle-to-grave' perspective. This transformation includes possibilities such as changing chemical reactions so that the yield of the desired product increases and fewer unwanted by-products are generated, and fewer harmful solvents are used.

(Sources: Schot (1992) and *Financial Times*, 13 March, 1991.)

Production

The focus should be on all aspects of the raw materials input both to manufacture and to use a product. Attention to the quantity and quality of raw materials input is obviously important, but concerns must also include the extraction methods used and the long-term reserves of natural resources. Referring back to the discussion on accounting and finance in chapter three, any consideration of inputted raw materials must take full account of their environmental costs, specifically, to what extent are we depleting non-renewable and non-substitutable, finite natural resources? It is important to use less, and less damaging, raw materials, including less energy and water. Simply stated, there should be a desire for a more efficient utilization of all resources.

Keywords in the management of the production process are traditionally efficiency and effectiveness, although 'sufficiency' is important because much unnecessary resource utilization and pollution are a result of less than optimum plant operations. An important management goal should be environmental improvements through the implementation of a 'less is more' philosophy, or a good housekeeping mentality. Special attention in this sub-section is given to energy efficiency, which is proving to be a significant business benefit for many organizations.

The use of energy in the Western world, for example, places immense strain on the world's environment and its natural resource base. The OECD countries account for 14 per cent of the world's population, 24 per cent of the land area, but over 80 per cent of the global energy use! Energy reduction can have benefits of increased returns due to cost savings in production combined with advantages of a better reputation derived from superior environmental performance. A recent study by the International Energy Agency, for instance, estimates that, with investments in known technologies, the expected global energy consumption at the millennium could be reduced by a quarter.

Not surprisingly, changes in resource inputs and associated processes are driven usually by the consideration of cost savings and risks concerning future availability. Energy efficiency, for instance, improved greatly after the oil price shocks of 1973 and 1979/1980; however, during the 1980s, the real price of energy decreased, and there was no longer a continuation of the market pressure to be more energy efficient. Hence, in considering energy efficiency, the progress, or lack of progress, is dependent on contemporary market prices which do not reflect the true environmental and external costs.

A significant component of many 1970's 'doom and gloom' scenarios was that the Earth would run out of energy resources in the near future. This argument has nearly disappeared, partly because of developments in alternative energy forms and partly because of discoveries of large, new reserves; for example, proven gas and oil reserves are nearly double their earlier estimates. However, increasing management pressures on costs mean that opportunities for energy efficiency should not be neglected, and the broader concerns about global warming mean that some reassessment of the energy mixes may be necessary at a national policy level. With cost reductions and positive environmental benefits, energy efficiency can be a clear example when business opportunity and business responsibility are simultaneously satisfied in managing our environment. Unless organizations have the necessary information to understand their current energy costs, it is unlikely that effective progress can be made, in terms of operating procedures and/or capital programmes. Not surprisingly, instant progress can be made through appropriate investment in new equipment to replace old equipment; evidence indicates, however, that often large savings can occur by greater operational housekeeping, monitoring facilities' performance.

Case: Reducing carbon dioxide emissions: A black hole?

Under compliance commitments related to the climate change convention at Rio de Janeiro's Earth Summit, it is necessary for Britain to go back to the 1990 levels of carbon dioxide emissions by the year 2000. Infuriating many of its industrial partners, the United States refused to set targets for cuts in carbon dioxide emissions; the country is the largest energy user and biggest producer of greenhouse gases, approximately 25 per cent of total carbon dioxide emission. Interestingly, the German Government supports the introduction of a carbon tax, but it has been attacked by business leaders because of the costs to them, which are estimated to be nearly DM 15 billion by the millennium.

Alternative, market– and regulatory-based policy options exist. However, some system of payments will be necessary. Unless the benefits of reducing global warming can be shared and are related to the countries' costs of action, incentives for free riders will remain. Simple flat targets are unlikely to be either effective or cost efficient, and there needs to be policies for agreement to concentrate on the more straightforward emission reductions in the short-term.

Costs of action rather than no action are usually discussed, with less attention being given to the benefits of action. OECD research estimates that stabilizing carbon dioxide emissions in OECD countries by the millennium to 1990 levels could result in an average loss of real income of 0.6 per cent over the period 1995 to 2050, as compared with no policy action. The comparable estimates for non-OECD

> countries of an average loss of real income of nearly 5 per cent
> highlights the envisaged difficulty of making progress.
>   Interestingly, the OECD research indicates that a general removal of
> distorting energy subsidies could have the twin benefits of reducing
> carbon dioxide emmissions and of increasing economic growth.

There remains much global potential to use renewable sources of power, such as energy supplies from solar power, hydroelectric power, tidal and wave power, geothermal, biomass, and wind, if the necessary planning and management are forthcoming. Across the world, there is now sufficient experience of different renewable energy sources to expect their contribution to continue to increase. Interestingly, often the emphasis of such so-called *Soft Energy Paths* is placed on their social and political, rather than their technical and economic, implications. Renewable sources are not a panacea. As the Worldwatch Institute argues,

> Hydroelectric power will not be truly renewable until the functions of flood control, irrigation, transportation, power production, tree planting, fisheries management and sanitation are co-ordinated with the overall goal of maintaining healthy and productive rivers.

Some commentators argue that a diversified energy policy that recognizes explicitly the varied options should provide a flexible, long-term basis. As a consequence, there is some cogency to the view that continued commitment to nuclear energy is appropriate. Nuclear power is a highly emotive issue, which has caused referenda in Austria and Denmark and civil disorder in a number of different countries. Contradictory claims have been made regarding the safety of nuclear power stations, disposal of nuclear waste, possible radiation effects, the chance of weapons falling into undesirable hands and so on. In Britain's privatization of electricity generation, the nuclear element was omitted because of uncertainties and complexities of adequate costings and estimates of possible liabilities.

Public confidence in the nuclear industry was eroded enormously and seemingly with some permanence by the accident at Three Mile Island in 1979 and by the 1986 Chernobyl disaster (and the much-publicized film, *The China Syndrome*). Since 1979, no new nuclear plants have been ordered in the United States. Today, the proportion of a country's electricity that is provided by nuclear generation varies enormously (see Table 5.1). Moreover, in aggregate, at a world level, this proportion is estimated to decrease in the future, because the growth in demand for electricity will be satisfied more by fossil fuels and renewable energy sources. Nuclear research, which is long-term with regard to commercial

Table 5.1   Proportion of Electricity
Generaged from Nuclear Power in
Different Countries

Country	Nuclear (%)
France	72.7
Belgium	59.3
Sweden	51.6
Hungary	48.4
South Korea	47.5
Switzerland	40.1
Taiwan	37.8
Spain	35.9
Bulgaria	33.9
Finland	33.3
Czechoslovakia (former)	28.6
Germany	27.6
Japan	23.8
United States	21.7
United Kingdom	20.6
Argentina	16.4
USSR (former)	12.6

(Source: Nukem, 1992.)

viability, is being focused on controlled fusion power, deriving energy by combining light atoms rather than splitting heavy ones. However, some industry commentators are concerned that the industry cannot remain viable without new construction and maintenance of new failities.

Case: Combined heat and power: A long neglected option in Britain

The majority of Britain's electricity is generated by using non-renewable fossil fuels, coal, gas and oil, which produce large quantities of greenhouse gases. Notwithstanding the safety and technical concerns about nuclear energy and continuing interest in renewable 'soft' energy paths, more attention should be given to well-developed and proven means of energy conservation. Energy conservation can be approached by two general, although not mutually exclusive, paths:

- reduce the level of demand for energy;
- improve the efficiency of its utilization.

Combined Heat and Power (CHP) plants increase the efficiency in the utilization of primary energy inputs, not only by generating electricity, but also by providing hot water for space and water heating, which is a major component of a nation's energy consumption for domestic, commercial and public service buildings. That is, it is the co-generation of electricity and hot water. The scale of the improvement in using the primary energy is generally from over 30 per cent to over 60 per cent.

It is interesting to note that, while less than 5 per cent of the UK's electricity is generated by CHP, the corresponding figure in Denmark is nearly 50 per cent. Indeed, across continental Europe, CHP schemes have a widespread and established incidence. Again, the financial feasibility of such CHP schemes is directly dependent on the existing pricing policies for energy. Moreover, while there are a range of small private schemes, restrictions on the sale of surplus electricity can be also another constraint on investment decisions.

Minimization of the use of dangerous components is another end goal in the life cycle concept, and environmentally damaging materials must, whenever possible, be substituted with environmentally neutral ones. CFCs are one obvious example that has been discussed already, and other harmful components include:

- toxins;
- phosphates in washing powders;
- nitrates and phosphates in fertilizers;
- chlorine;
- chemical pesticides.

For instance, chemical pesticides, especially the chlorinated hydrocarbon pesticides such as DDT, have been known for years to have detrimental, long-term effects on the environment, wildlife and on human beings. Alternatives to chemical pesticides include biological control using natural predators.

From a general management sense, large-scale storage of raw materials, semi-processed materials and finished products can be thought as inefficient, unnecesarily tying up financial resources. Again, more than one aspect must be taken into consideration, because storage relates to not only raw materials, semi-processed materials and the product itself, but also the environmental risk involved in the storage of hazardous waste resulting from the production process. Caution with dangerous, hazardous materials and waste is important for secure, environmental protection; such waste is often expensive and difficult to store or dispose of properly. Risk arises as a result of the types and quantities of materials held in store and also the mode of storage. Environmental risk tends to rise if dangerous materials are stored in bulk or under pressure.

Case: Fuji Film: Do not 'carp' about the environment, act!

While better known as a photographic organization, the 'green dragon' fighting the 'yellow elephant' of Kodak, today, Fuji Film has a growing involvement in and broadening corporate philosophy towards imaging and information.

Fuji Film's Ashigara factory is sited in a naturally beautiful location in the foothills of Mt. Fuji, with the green expanse of the Sakawa plains to the north and east and the Hakone mountain range to the south. Its main production item is photosensitized products and associated materials. The critical location force behind the selection of a site for film manufacture is an abundant supply of water, the quality of which affects directly the quality of the finished product.

Daily, the Ashigara factory consumes approximately 60,000 tons of water, with about half of it being used for washing the air introduced into production areas. For the film industry, it is required that the air is washed at a constant humidity and temperature throughout the year. Water is also used in large quantities to make emulsions and in other production processes. It is, therefore, not surprising that Fuji Film are extremely environmentally conscious to maintain the factory's supply of water.

To help protect the environment, Fuji Film uses the latest technologies to avoid harmful production discharges and maintains an active tree planting programme at Ashigara. An anti-pollution committee was established at Ashigara in 1965, and, in 1970, an environmental group was established in all its production facilities. For its worldwide activities, Fuji Film has an Environmental Charter which states,

- We shall do our best to safeguard the environment and uphold water and air qualities.
- We shall exercise constant vigilance that our products are not injurious to health and safety.
- We shall strive, in every sphere of corporate activity, to conserve energy and resources.
- We shall make concerted efforts to study, propose, plan and implement programmes that will advance environmental causes.

In 1983, Fuji Film created a Fuji Film Green Fund by placing one billion yen in a public trust to assist selected nature projects and ecological research.

At Ashigara, all the liquid wastes, including factory effluent and restaurant sewage, are piped to the liquid treatment plant within the factory compound and restored to clean water by a range of processes, including activated sludge treatment, coagulation, skimming, filtering, and ozone treatment (see Figure 5.7). Visual surveillance of liquid storage is used to ensure no polluting fluid, whether chemical, effluent or fuel, is allowed to escape into the ground.

Water conservation is also important. Various techniques have been developed to recycle industrial water, up to nine times. Although Ashigara now produces four times more products than it did fifteen years ago, it consumes a fifth less water.

As a visual demonstration of the factory's water quality before it is discharged back into the outside water system, a 5,000 square metre pond is the home for about 20,000 coloured carp and a place for employees to relax.

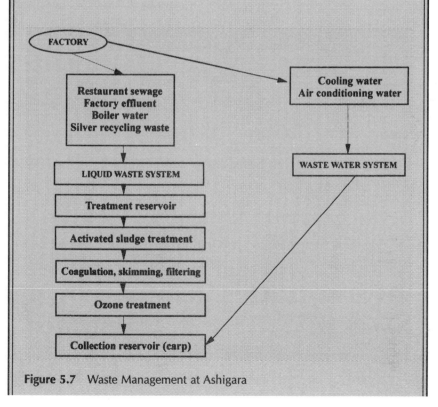

Figure 5.7 Waste Management at Ashigara

As well as being environmentally conscious regarding its production and operations management, Fuji Film has also considered its products' impacts on the environment. In 1986, for example, they launched a disposable camera into the Japanese market, and, in 1990, they introduced an effective disposal system through component separation and the reuse of the flash bulbs.

Fuji Film has a saying to illustrate its environmental commitment,

Preserving more than just memories.

Distribution

Transport is a major contributor to pollution and finite resource depletion. The issue of distribution is mostly related to energy use in transportation (see also the discussion in chapter four). For a complete view, both the transportation of inputs to and outputs from the production processing facilities must be examined. An efficient supply-distribution system is closely linked to plant location and size, the available transport infrastructure, and the characteristics of the local markets' demand.

Logistics is the spatial or geographical dimension of the flows of materials and products. In the 1970s and 1980s, functional integration within organizations occurred to help to overcome suboptimality arising from individual function's myopic perspective that was concerned primarily with the performance of their domain of responsibility rather than performance of the whole organization. For the 1990s, such integration will be driven increasingly beyond the single organization through supply and waste chain management, linking suppliers, customers and third parties. The growing focus on such collaboration and partnerships is beginning to replace the more traditional and confrontational (power-based) buyer-supplier relationships. Against this changing backcloth, there could be direct and indirect environmental benefits. Just-in-time management is an inventory control and customer service philosophy, which reduces unnecessarily high inventory levels, whether supplies of raw materials, work in progress, or stocks of finished products.

In a simple, idealized form, it is possible to differentiate between:

- inputs – raw materials supply;
- throughput – manufacturing processes;
- output – distribution.

MRPII, or manufacturing resource planning, has been an established logistics tool for the inputs – throughputs interface; more recently, the application of similar principles to the throughput – outputs interface has led to the development of another tool, DRP (Distribution Resource Planning). Necessary scheduling downstream arises because, in many supply and waste chains, a complex distribution network exists with associated costs of say about 20 per cent of the total costs. As well as environmental benefits, the types of business benefits can include:

- greater productivity;
- reduced operating costs;
- enhanced customer targeting;
- better deployment of people to complete added value tasks.

At present, in the United Kingdom as indicated by Table 5.2, road transport is the dominant mode for freight transportation.

Table 5.2   Britain's Modal Split for Goods Transportation (1990)

	Tonne kilometres (thousand million)	*Tonnes (million)*
Road	136.2	1,749
Rail	15.8	141

(Source: Department of Transport.)

An early discussion of *Transport and the Environment* by Clifford Sharp and Tony Jennings (1976) presented different reasons for the attraction of roads, including:

- cost;
- flexibility;
- reliability;
- speed.

Obviously these characteristics are relative to other, competing modes of available transportation, and, for Britain, some people have argued the lack of investment in public transportation, particularly the railways, has been misguided. The privatization of British Rail remains unclear, although there have been suggestions that subsidies for freight transportation will exist.

The Single European Market, which was created on 1 January, 1993, is likely, over time, to have significant effects, not only on modal split decisions, but also, more fundamentally, on the geographical perspective of the logistics function, especially as the trend to use third-party hauliers is envisaged to continue. Increased transborder transportation of products can be expected. Interestingly, shortly after being appointed EC Environment Commissioner in July 1992, the Belgium Carel Van Miert made it clear that transport must be brought into line with other EC objectives on the environment. One of his aims is to make road haulage more expensive, as well as promoting canals, railways and shipping.

Information and communications technologies are assisting distribution by improving existing practices and by permitting completely new ways of working. Given the enormous impacts that transportation will continue to have on the environment, it is noted that business and environmental benefits are accruing because systems are starting to enhance transportation efficiency and speed up the flow of materials and products through a greater mangement focus on the effective use of time spent 'on the road'. Intelligent Vehicle Highway Systems is a generic term that describes several technologies for advanced traffic management systems, including automated vehicle location, speed monitoring, incident detection (via inferences from speed data), route planning and so on.

Use

The ways in which products are used or consumed can have adverse effects on the environment. Indeed, when considering their entire life cycle, for many products, it is their use that has the most significant effect (see, for example, Figure 5.3). For completeness, it is sufficient to reference here a number of issues:

- the direct linkage with product design;
- the quality characteristics of product reliability and durability;
- the possibilities of product substitutability;
- the level of consumers' environmental consciousness;
- the availability of comprehensible product information.

Waste management

In most production processes, there are two outputs:

- the product;
- the waste.

They must both be disposed of in the safest and most environmentally friendly way possible. It can be argued that waste is one indication of organizational inefficiency (although, to some extent in all organizations, inefficiency is partly historical and institutionalized). Different types of waste can be broadly identified:

- industrial waste, which is produced as a direct result of manufacturing processes, including hazardous effluents and emissions;
- domestic waste, including packaging, which is generated as a by-product in the consumption and use of products and services;
- product disposal at the end of their apparently useful life.

Business benefits through cost savings exist from waste minimization and reuse. For example, in 1988, the Dow Chemical company introduced its Waste Reduction Always Pays (WRAP) programme in the United Kingdom from the US. One of the first areas to implement the policy was the company's weedkiller production plant in King's Lynn, East Anglia. It is estimated, that more that £500,000 a year has been saved at the plant. As well as direct business benefits, given the visibility of waste, enhanced waste management can assist the corporate image of an organization.

In the 1990 UK Environmental Protection Act, the 'Duty of Care' requirements, have meant that, since 1 April, 1992, a producer has become responsible for its waste, including the illegal disposal by other organizations. Unlimited fines are possible for breaking this law. The British Government has issued practical guidance on the Duty of Care, indicating what is reasonable in different situations. Maintenance of detailed records of all transfers of wastes is required, and these must be available for external inspection.

As to be expected, the aggregate level of waste tends to fluctuate with recession or economic upsurge. However, since World War II, with the rise in Western consumerism, waste is now a major concern. In general, there are differences in the stages of development for waste management for industry and domestic waste. Although there is no reason for complacency, some claim that industry is doing far better; developments are still at a very early stage for domestic waste management. Visible illustrations of the waste crisis are urban waste sites. For New York City, Fresh Kills landfill on Staten Island receives 44 million pounds of rubbish each day, which is making it one of the highest points on the eastern seaboard. The amount of industrial and domestic waste produced worldwide has increased steadily over the recent years, particularly in the twenty-four industrialized countries of the OECD. In 1990, for example, Western Europe, the US, Canada, Japan, Australia and New Zealand produced a total of 9 billion tonnes of waste. This total included:

- 420 million tonnes of household waste;
- 1.5 billion tonnes of industrial waste (of which 300 million tonnes were hazardous);
- 7 billion tonnes from activities such as energy production, agriculture, mining, and sewage disposal.

These totals amount to about 10 tonnes per person per year! The total weight of UK waste rose by 3 per cent in 1990 to a total of 516 million tonnes.

Case: International hazardous waste trade: 'Waste imperialism'

In 1991, approximately 200 tonnes a week of hazardous waste was going into Argentina from Europe and the US. In addition, Argentinean businessmen were planning to import about 250,000 tonnes of plastics a year for incineration and land dumping. Moreover, they are finding a market for their services; there are few countries, it seems, which are not prepared to send waste to Buenos Aires.

For Argentina, it is also possible to read Guatemala, Poland, Romania, Chile or a dozen other countries that are now in the receipt of a massive export drive by more environmentally aware countries in the world. The waste trade is seen as an increasingly significant wealth earner for the 'developing' countries, and, therefore, it is welcomed by them. For example, the US, Germany, Holland, Switzerland and the Nordic countries, which are introducing ever tighter environmental legislation, are now exporting millions of tonnes a year of hazardous wastes to countries with weak laws and inadequate administrations that are unable to effectively manage the waste. Many countries, which willingly or inadvertently accept the toxic waste for landfill or incineration, are ill-equipped to handle the waste safely. Unfortunately, the incidence of serious environmental health and long-term ecological problems is known to be increasing.

Interestingly, under the so-called Bamako Convention, African countries have signed a treaty to agree not to import any hazardous waste. Under Austrian law, waste exporters are expected to know that their waste will be disposed of as effectively as if it was completed domestically.

Critics report that waste plants in developing countries are always poorly equipped, with low environmental health standards. Andreas Bernstorff of Greenpeace in Germany has stated recently that,

In 99.99 per cent of the cases, it is 100 times safer, ecologically, for the rich countries to treat hazardous waste 'at home' where technological standards are much higher. No developing countries insist on the cleanest technologies. Very few have any idea of the long or short term dangers of what they are handling.

Jim Valette, Bernstorff's opposite number in New York, has indicated that,

We have investigated the trade throughout the world, and nowhere, ever, have we found clean plants. In every case where waste is processed there are definable human health impacts.

It can be argued that environmental pressures in Europe and the US have only shifted the problems on to the weakest nations. The developed countries' new standards of cleanliness are leading directly to waste imperialism. The primary goal for managing our environment of waste avoidance, in both consumption and production, is not being achieved. For instance, Germany is the largest exporter, and was known to be shipping more than half a million tonnes to some fifty, different countries in 1991. It is perhaps no accident that Germany also has some of the toughest environmental laws in the world. Close behind in terms of tonnes per person come the Dutch, shipping more than 250,000 tonnes in 1988, the US with 141,000 tonnes, and Switzerland and Austria with 200,000 tonnes between them.

As regulations on disposal are tightened in Europe, pollution problems are exacerbated elsewhere. Through market mechanisms, skewed price differentials are acting as an impediment to pollution control in developing countries. The costs of disposals in the more environmentally aware developed countries are rapidly increasing as more and stricter legislation is introduced. Many companies are finding it unprofitable to take their waste to new European incinerators. Consequently, there are also disincentives for waste companies to build more incinerators, at a time when so much trade is going abroad. The cost differences are not only between developed and developing countries; it is about six times more expensive to burn solvents and paint waste in Germany than it is in France.

Eastern Europe is becoming the most attractive destination target of Western European countries. In the last two years, more than 70,000 tonnes of waste has been shipped into Poland. However, a waste import ban, which was hardly noticed when it was first introduced in 1989, is beginning to take effect. Yet, 23 million tonnes of hazardous waste was offered to the country between 1989 and 1991. Ships are now being stopped, trucks are now being sent back, but it is reported that a lot of waste still makes its way through administrative loopholes and a lack of expert scrutiny at customs. Romania and some of the

Baltic states, such as Estonia and Lithuania, are now targeted to become the dustbins of Europe. Germany proposes to export two and three million tonnes of shredded residues from cars for use in land reclamation in Romania. Romania has been approached by German, Italian and Austrian companies with offers of free construction of toxic waste incinerators in return for exclusive use of half of the plants' capacity for the first ten years of their operation.

In taking a global view of managing our environment, it should be appreciated that the marginal improvements gained from a particular level of expenditure may be greater in developing countries, rather than in developed countries. Both businesses and governments must recognize their responsibilities, as well as understanding the basic driving forces behind the current trade opportunities.

(Source: adapted from *The Guardian* and other reports.)

Given the increase in the quantities of waste and much greater concerns for the quality of its management, a range of organizations have seen the new business opportunities and are now providing waste management services, including a wide range of specialities such as hazardous waste disposal, recycling, environmental engineering, and water and air treatment. In fact, through joint ventures, mergers and acquisitions, company 'spinouts', and so on, the waste management industry is now beginning to mature in many countries. Moreover, the nature of their services is also changing, going up the waste management hierarchy of options. Waste Management Inc., for example, is one of the world's most well-known environmental services companies. Over the last decade and a half, the company has exhibited profitable growth; since 1980 revenues have increased at an annual rate of 27 per cent and earnings at 29 per cent, and, today, the company is approaching a $10 billion a year business. Waste Management Inc.'s business environment is changing rapidly. In the United States, where landfills are becoming increasingly unpopular, and where waste minimization and recycling are gaining popularity, Waste Management Inc. is changing strategy. Landfills and the handling of solid waste will still be important for the company, but from concentrating on operating landfills all over the US, growth in the future is expected to come from dealing with waste at source.

Waste management options include:

- waste minimization at source;
- reuse;
- recycling (including composting);

● incineration or waste combustion, with energy recovery;
● landfilling.

This list is a hierarchy of preferences, with source reduction is the ultimate goal in a cradle-to-grave perspective. The most effective means of solving the waste crisis is to prevent it from being created in the first place. However, before this aim can become a reality, a significant shift in mentality and political priorities towards reduction and recycling is necessary. Recycling should not be prioritized over source reduction, which is the only alternative that slows down the depletion of available resources. A slogan, which is used sometimes, is:

Reduce, Reuse and Recycle.

Failure to reuse or recycle is wasteful of scarce resources, and an unnecessary cause of adverse environmental impacts. For instance, the United Nations Environmental Programme (1991) *Data Report* not only indicates enormous variations in countries' recycling (and thereby gives some indication of the order of missed potential), but also highlights the decreasing levels of recycling in many countries.

Recycling has been promoted significantly in recent years. For instance, according to *This Common Heritance*, which was published in September 1990, the British Government's plans for the end of the century are to recycle half of the recyclable (which is about quarter of all) household waste. This target compares to the current level of under 10 per cent, and concerns about the realism of the target have been expressed. Recycling is not a viable solution unless demand for the recycled products exists. Similarly, there must be consistent and reliable sources of supply and associated processing facilities. Whether it is because of detrimental recessionary impacts on the levels of demand for many recycled products and/or their increased supply because of the growth of recycling capacity, excess supply of recycled (glass, paper and plastics) materials is meaning that many of these products remain unsold. This situation exists in many parts of the world. In Britain, the paper recycling industry now has an excess supply of recycled paper, and the clear implications of the lack of demand is a reduction in the country's recycling capacity. In New York, as a further market distortion, the city government pays recyclers to continue their activities because it is their cheapest form of waste mangement, even though there is not a sufficient market for the recycled materials.

Currently, more than 90 per cent of all controlled waste in the UK is disposed in landfills, and UK waste policy is still based on disposal into landfill sites. With regard to incineration, the UK remains behind the rest of Europe, and currently mostly hazardous waste is incinerated. Trade-offs in waste disposal between landfills and incineration is a topic much debated by both the waste management industry and environmentalists. The latter focus on the possible dangers of toxic emissions from burning waste; the former argue that, without proper incineration facilities, which are much safer now than a decade ago, Britain will face greater environmental damage through improper disposal of waste or accidents involving dangerous waste in long-term storage. In any event, more attention should be given to a risk assessment of these alternatives.

In any case, the emphasis on incineration and landfills allows continuation of waste production rather than minimization. For instance, the situation in the UK contrasts with the situation in the US where all waste disposal options are available in a more integrated waste management system. Waste minimization is the main solution, followed by recycling and incineration; landfills in the US are considered the least preferable disposal route. For the future, many countries are rapidly running out of landfill capacity, and tipping fees are increasing as a consequence. In addition the understandable 'NIMBY' opposition is slowing the opening of new landfill sites, and authorities are being forced to change priorities and look at alternative solutions. German officials, for example, estimate that within ten years, the country's landfill capacity will be exhausted, and environmental problems have increased after the unification with severe clean up needed in the former East Germany. Levies on waste have been proposed, with highest ones on toxic waste. The carrot is clear; companies which invest in waste reduction will pay lower charges. German industry is not impressed claiming that such a waste tax will be too costly resulting in a competitive disadvantage.

For Europe, the EC has put forward far-reaching proposals to harmonize environmental laws and to reduce the 50 million tonnes of packaging waste produced in the Community each year (see also chapter four). Its hierarchy of preferred solutions is: minimization of the packaging used in particular products; followed by recovery and recycling; with final disposal in landfill sites remaining last resort. The directive sets the seemingly staggering 'aspirational' target that within 10 years of implementation 90 per cent of all packaging waste by weight should not reach the waste stream and 60 per cent of each packaging material should be recycled (see Figure 5.8). The EC expects that, within five years of implementation, the member states will have appropriate systems to collect packaging waste from the consumer, enabling them to ensure that it is effectively reused or 'recovered'. The concept of recovery is a loose one, much to the packaging industry's relief, because it allows for the value of any product to be recovered by the most efficient means possible; for example, it will be considered equally valid either to recover the energy from a product by burning it and utilizing the heat or to compost products for agricultural use.

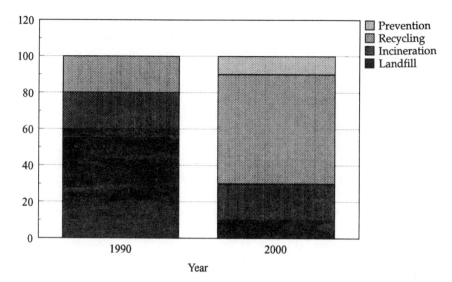

Figure 5.8   EC Packaging Waste Management Objectives for the Millennium

Case: The 'greening' of car manufacturers

As the environmental concerns have now shifted from considerations of resource shortages to broader aspects of whether our planet can cope with the current pattern of activities with their depletion and degradation of resources and their production of pollution and waste, the motor car has continued to illustrate the difficulties. The scale of car ownership and usage has been underestimated continually. While the car's implications for finite resources have been debated for decades, the growing interest in a car's entire life from design, through production and use, to disposal has become of increasing significance.

In this relatively long case, some of the environmental issues concerned with the design, production, use and disposal of the motor car are explored, with the specific orientation being as to how these characteristics are now being promoted by car manufacturers through advertising (see also the earlier discussions in chapter three and four). No attempt is made to be comprehensive, in terms of either all car manufacturers and their different models or the large range of complex and interrelated issues about our motorized society and our physical environment. The discussion is essentially a British consumer's perspective, although some international references are made as signals of what could/might occur in the future.

As with all production and operations processes, there are adverse environmental impacts resulting from the manufacture of cars, particularly (air and water) pollution and waste. Although progress has been made by say the increasing replacement of solvent-based paints by water-based paints, the major change has come from the broader 'life cycle' perspective. Rolls Royce, for example, with petrol consumption averaging 18 miles per gallon, claims that, because its cars generally last about six times longer than the 'normal' family car, its overall energy efficiency is significantly greater when manufacturing energy consumption is also included in any analysis. With a cradle-to-grave outlook, the importance of disassembly and recycling are also highlighted, as well as the need for explicit linkages from design, through manufacture, use and repair, to disposal and recycling. Volkswagen, for instance has a '3V' environmental policy: Vermeidern, Verringern, and Verwerten (Prevention, Reduction and Recycling).

Appropriate design can lead to more manageable disposal and/or recycling, especially if the cars are designed with fewer, more recyclable materials. Approximately 10 per cent of cars are disposed of each year. It is difficult to exaggerate the complexities of recycling; cars are composed of approximately twenty thousand parts, which are made from a variety of materials. ID numbers on individual components can assist the necessary identification of recyclable and non-recyclable components.

The manufacturers that are taking a responsible lead for car waste management have recognized the business opportunities from such action. Others, however, may have to comply with future legislation; in Germany, for example, legislation to require the manufacturers to be more responsible is already envisaged.

More specifically, much research is being undertaken into new engine technologies, ranging from extensions into direct injection diesel engines to greater use of ceramics and plastics. So-called 'lean-burn' engines can simultaneously enhance fuel efficiency and reduce exhaust emissions. Real progress is being made generally by the substitution of metals by plastics, with direct benefits of improved fuel efficiency because of reduced weight and longer material life because of erosion resistance properties.

The issue of alternative fuels is dependent ultimately on when they become commercially viable; the alternative fuel systems have volume and/or weight problems when compared to the conventional systems. Engine technology remains one of the most difficult problems, with no real commercial solutions likely on a short- or medium-term horizon. Petrol's energy efficiency remains at only approximately one third. Vauxhall's promotion of its Cavalier attempts to overcome consumers' perceptions of diesel as being dirty and smelly. Methanol, for example, could have some attractions if it can be used effectively in traditional engines; solar power may have some possibilities in certain regions of the world. Compressed natural gas and hydrogen are amongst the fuels also being explored. The electric car, with the attractive features of no engine emissions and low noise, has been on the horizon for some years. They have been used for a number of years for golf buggies, milk deliveries and fork lift trucks, but they do not possess currently the required range and speed capabilities for a reasonable cost and durability. Fiat have electric Pandas, Eletra, in Italy; a Danish company, CityCom, makes electric 'personal commuter vehicles'. In 1994, electric family cars are expected from Citroën, Fiat, Ford and Volkswagen, but they are also expected to be relatively expensive. However, it should be considered to what extent electric cars would only move the pollution source from cars to power stations.

Volvo's environmental concept car (ECC) is a first prototype to combine gas turbine with battery power, seeking to enhance both energy efficiency and cleanliness. In cities, the batteries will provide the power with zero emissions, and, out-of-town, the gas turbine will emulate the performance and range of a petrol car (with lower emission levels).

The pressure from the car manufacturers, because of their large stocks of unsold cars, has meant a temporary reprieve from a European Commission requirement that, from 1 January 1993, all new petrol-engined cars should have had a catalytic converter and use unleaded petrol (but petrol companies envisage supplies of leaded petrol will continue for at least a decade). This specification will lead to many smaller models having electronic fuel injection to ensure the efficient operation of the catalytic converter. As a cautionary note on technological advances; it is often assumed that the remedies work all the time. Tests undertaken in cooler climates indicated initially some inefficiencies in converters' operation, with both increased fuel consumption and greater exhaust emissions!

It is conventional to view waste management as a series of hierarchical levels. For example in the car manufacturing industry, Toyota has a model of five Rs based on:

- Refine;
- Reduce;
- Reuse;
- Recycle;
- Recover.

Similarly, BMW have a so-called 'cascade' model:

- disassemble for reuse:
  - as a cautionary note, it would be misleading to believe all reclamation of components is a recent phenomenon; reconditioning of relatively high value parts, such as different body sections, engines, gear boxes and radiators, has been a profitable activity for car scrappers.
- use old parts for making new parts:
  - damaged or plastic bumpers from the BMW 3-series, for example, are now collected, shredded and returned to Germany for reuse as boot linings, floor mats, and so on.
- extract chemical components from parts:
  - for example, with a specialist task, the removal of precious metals, platinum and rhodium, from catalytic converters can yield up to 80 per cent reclamation. For batteries, the acid and lead are removed for chemical recycling, and their plastic cases are melted for reuse in lower grade components.
- produce energy:
  - from the remainder, where possible, generate usable energy.

Co-operation, across the industry's supply and waste chain, has increased significantly. Concerns, for instance, for the environment and safety led to technological co-operation under Prometheus (Programme for European Traffic of Highest Efficiency and Unprecedented Safety), a consortium of manufacturers that included BMW, Jaguar, Lucas, Mercedes-Benz and Rolls Royce. As well as technological co-operation, there is increasing operational integration across the industry's supply and waste chain; for example, suppliers are now involved in reconditioning certain components.

Business process redesign and business scope extension are really required (see also chapter two). In the traditional wrecking and shredding process, after the saleable parts are removed, the car is pressed, with a shredder taking out non-ferrous metal. The waste is about a quarter the weight of the original car, but it is contaminated by glass, fluids, plastics, rubber, and so on. Recently in the UK, BMW launched a two part recycling scheme with a co-ordinated national network to disassemble and reprocess or dispose of cars and their components. Clean disposal of batteries, brake fluid and motor oil is a significant initiative.

While not concerned specifically with production and operations management, given the discussion of cradle-to-grave management, it is appropriate to now briefly discuss the use of the car and its environmental impacts. A driver's capabilities directly affect the environmental pressures, not only in terms of unnecessary fuel consumption and noise, but also in terms of accidents, maintenance, commuting and so on.

A car's operational efficiency, and, indeed, its effective life, can be enhanced greatly by regular servicing and repair. Out-of-tune cars waste fuel and also emit more toxic gases than should normally be expected. Car exhausts are a major source of nitrogen oxides, creating atmospheric acidification.

In comparison with the United States, in Britain, the introduction of unleaded fuel was much later and no transitional arrangements for complete replacement of leaded fuel have been formulated. Britain has been opposed to banning this poisonous heavy metal, attempting to encourage the use of lead-free fuel through preferential tax levels. In early 1988, Toyota (GB) brought to the market the first car in Britain to be fitted with a catalytic converter - the Celica GT-Four. A three-way catalyser meant that only unleaded fuel could be used. At this time, the availability of unleaded fuel was very rare. A five litre can was fitted in the boot to provide additional, spare capacity! To help alleviate customer concerns regarding the ability to obtain unleaded fuel, in a joint publicity exercise, Toyota with Shell and *The Observer*, a Celica GT-Four was driven from John O'Groats to Lands End, never with more than 10 litres of unleaded fuel in the tank and only able to refuel at Shell stations. The car completed the journey without running out of fuel!

In terms of energy efficiency, again on the John O'Groats to Lands End run (and back to Scotland), an Audi 100 TDI was driven recently further on a single tank of fuel than any other car; on 17.62 gallons of fuel, it averaged 75.94 miles per gallon. It is interesting to note that, while large increases in fuel efficiency have occurred over the last twenty years, of the order of a quarter to a third improvement, the potential for improvement is shown by a comparison of the variations in mean miles per gallon statistics for different countries. Volkswagen is experimenting with an automatic stop-start system that will save fuel at traffic lights specifically and in congestion generally.

Car manufacturers, especially European manufacturers of luxury cars, have paid more than $233 million in fines since 1985 because of their failure to satisfy US fuel efficiency standards. Violation of the Corporate Average Fuel Economy (CAFE) standards, currently 27.5 mile per gallon, means that car manufacturers have to pay fines, but the system permits the manufacturers to keep their fuel inefficient cars in the US market. Each car manufacturer's fleet must, on average, meet the standard, and, therefore, relatively efficient models can be used to offset the relatively inefficient. The current fine is $5 for each tenth of a mile under the standard multiplied by the number of cars sold in the

> US in the year. Mercedes-Benz and BMW are the two companies that
> have paid the most fines. Since 1991, in the US, there has also been a
> luxury tax on cars costing more than $30,000.

Concluding comments

Throughout this book, our line of argument is that business-led voluntary action, implementing sound environmental performance throughout the organization, is preferable to reaction to legislation. In this way, businesses are in charge of their own destiny and in charge of shaping the future benefits of long-term profitability and viability. However, a framework of properly enforced legislation is also required to set minimum standards. Indeed, it may be difficult to introduce fundamental changes in production processes without a regulatory imperative, because the requirements are most often needed in declining, mature industries.

A move towards implementing the cradle-to-grave management philosophy will mean fundamental changes in the production and operations management function. The narrowly defined, linear link between input and output is no longer sufficient. A holistic view, investigating the real environmental impacts, is an imperative. For such a vision to be implemented successfully, changes in interfunctional relations within organizations towards closer collaboration both on the operational and strategic levels are necessary.

The winners of the tough competition in the 1980s were the companies incorporating total quality management in production and operational processes (mainly Japanese companies), and it can be argued that the winners in the decades to follow will be the forward-looking companies incorporating the 'life cycle' perspective in their production and operations for effective management of our environment.

Case: Freedom to pollute: Just between you and me

In a much-publicized, leaked memorandum which appeared in *The Economist* in February, 1992, the Chief Economist at The World Bank, Lawrence Summers asked: 'Shouldn't The World Bank be encouraging more migration of the dirty industries to the less developed countries?'

Three reasons were present for this controversial proposal:

- with costs of pollution dependent on earnings lost through death and injury, select locations for polluting industries and hazardous waste disposal in the lowest wage countries;
- with the costs of pollution rising disproportionately as it increases, select the world's cleanest locations rather than make dirty ones still dirtier;
- with richer people generally valuing the environment more, the costs of pollution will decline if they are moved from 'rich' to 'poor' countries.

Notwithstanding the alarm of suggesting life in some countries is worth more than in other countries, Summers does raise an important issue that must be addressed.

To what extent should countries be able to establish different environmental standards?

The long-run economic advantages for business of cradle-to-grave management are becoming evident. However, many companies still need to be informed and convinced, and the issue is more complex than say tactical changes to an advertising campaign. All businesses cannot adopt a cradle-to-grave approach easily. It is obviously more straight forward to introduce it in new, fast-growing industries that have the opportunity and flexibility to build in new technology and new ways of thinking in production. Unfortunately, the big polluters are often to be found in traditional industries where slow growth discourages new investments and makes it more difficult to change conventional manufacturing processes. Moreover, in practice, it is important to differentiate between large and small and medium-sized organizations, not only because obviously the scale of their environmental impacts vary, but also because their scope to act varies. There is a real need to consider small and medium- sized enterprises, which are the majority of organizations, both as individual organizations and as a component of the industry's supply and waste chain.

These important barriers to progress can, in part, be overcome through pooling knowledge, spreading information, and collaborating, which can spur new ideas, innovation and increased investments in sustainable production based on clean technologies and a 'cradle-to-grave' perspective. With the life cycle concept, the foundation for the virtuous circle has been created which can replace the vicious circle of our current industrial societies.

Joseph Stilwell and his colleagues (1991), for instance, outline a conceptual model for contemporary patterns of industrial activity that can evolve to a structure they call 'industrial ecology' (see Figure 5.9).

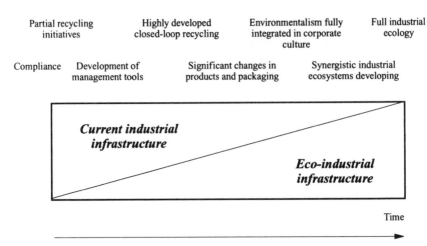

| Partial recycling initiatives | Highly developed closed-loop recycling | Environmentalism fully integrated in corporate culture | Full industrial ecology |

| Compliance | Development of management tools | Significant changes in products and packaging | Synergistic industrial ecosystems developing |

Figure 5.9   Timeline of Evolving Corporate Environmental Response
(Source: adapted from Stilwell, Canty, Kopf and Montrone, 1991, page 165.)

In their model, Joseph Stilwell and his colleagues view the natural environment as the model for the solution, rather than the conventional focus of the problem. They argue,

> The most important feature of the natural global ecosystem is that all outputs are inputs somewhere else. There is no such thing as 'waste' in the sense of something without use or value. Materials and energy are continually circulated and transformed. From the death and decay of one species come the life-giving nutrients for another. Thus, all parts of the system act independently, but are meshed cooperatively.
>
>   These are some of the important features of an industrial ecosystem:
>
> ● compatibility with the natural system;
> ● maximum internal reuse of materials and energy;
> ● selection of processes with reusable waste;
> ● extensive interconnection among companies and industries;
> ● sustainable rates of natural resource use;
> ● waste intensity matched to natural process cycle capacity.
>
> (Stilwell, Canty, Kopf and Montrone, 1991, page 165.)

Review questions

1. Using appropriate examples, describe what you understand to be cradle-to-grave management.
2. For a particular car manufacturer, explore the extent of its responsibility to the environment.
3. Discuss the management issues of energy efficiency.

Study questions

1. Examine the general topic of total quality management, and, using an organization that you know, consider the extent that it is consistent with improved environmental management.
2. Explore the current state of the international trade in waste.
3. Describe how you would attempt to promote energy efficiency in industry, and develop the outline for a national Energy Policy.

Further reading

No book or paper with an environmental orientation for production and operations management is recommended. However, for more detailed understanding of this important, albeit neglected, functional area, a number of volumes are suggested.

Hayes, Robert H. and Wheelwright, Steven C. (1984) *Restoring Our Competitive Edge*, John Wiley and Sons, New York.
Many countries are permitting their traditional manufacturing base to disappear, forever. In this excellent volume, Robert Hayes and Steven Wheelwright demonstrate how world-class organizations can compete through manaufacturing. Four critical activities are highlighted:

- developing appropriate production facilities and managing their evolution;
- choosing equipment and management systems appropriate to those facilities;
- establishing supplier relationships to provide them with parts and services;
- encouraging continual improvement in their performance.

Hill, Terry (1989) *Manufacturing Strategy*, Macmillan, London.
A practical, short book that emphasizes the need for a strategic perspective for manufacturing. An approach is developed that should help to indicate to managers the implications of the corporate marketing and finance decisions for their manufacturing processes and infrastructures.

Lockyer, Keith, Muhlemann, Alan and Oakland, John (1988) *Production and Operations Management*, Pitman, London.
In its fifth edition, this text provides a comprehensive introduction to production and operations management.

Schonberger, Richard J. (1982) *Japanese Manufacturing Techniques*, Free Press, New York.

An early, comprehensive and clear description of Japanese manufacturing practices. Nine lessons are presented:

- management technology is a highly transportable commodity;
- just-in-time production exposes problems otherwise hidden by excess inventories and staff;
- quality begins with production, and requires a company-wide 'habit of improvement';
- culture is no obstacle; techniques can change behaviour;
- simplify, and goods will flow like water;
- flexibility opens doors;
- travel light and make numerous trips;
- more self-improvement, fewer programmes, less specialist intervention;
- simplicity is the natural state.

6 Information resources management

The study of global environmental change will involve the collection and interpretation of data collected from the atmosphere and ocean, on the ground, from space, in laboratories and within societies. Such data will form a vital national and international resource which will enable the processes and impacts of change to be detected, studied and predicted, and responses planned. These data are a crucial . . . resource, not merely an end product of observation.

(UK Inter-Agency Committee on Global Environmental Change, 1991, page 15.)

Introduction

Notwithstanding the rapid technological developments of the 1970's and 1980's, it is a sobering thought that the majority of developments in the information and communications technologies lie in the future. Moreover, the vast majority of organizations have been unable to exploit the business potential of the existing technologies. In this chapter, the phrase 'information and communications technologies' is used, because, given the recent technological convergence of computing and communications, it better captures the nature of this technology platform than the traditional term 'information technology'. John Beaumont and Ewan Sutherland's (1992) discussion of *Information Resources Management* illustrates the necessary perspective that information is a strategic resource for organizations that requires management.

For managing our environment, the usefulness and relevance of information and communications technologies should be demonstrated through their applications in terms of:

● providing information as an input into management decision-making;

● offering a technology platform as an infrastructure that creates opportunities for new ways of doing business.

The usual assumption is better and more appropriate information should lead to better decision-making; as stressed already, for managing our environment, evaluation of alternative options, as well as the monitoring of any implementations, will require the effective handling of data. New ways of doing business that can have environmental benefits include, for example, home shopping as discussed briefly in chapter four.

The design, specification and implementation of any information system whether computer based or not, must be driven by an understanding of the information needs of the management. In order to achieve that understanding, managers and organizations must know:

● what information they need;
● what information they have access to;
● what information they can use effectively.

In this chapter, a brief, non-technical overview of computer-based information systems is provided, which is developed, in section three, by linking the discussion to management decision-making. The central, simple argument is that the nature of environmental decisions should drive the information requirements, and that investments in information and communications technologies should not be seen as for merely automating existing processes and procedures. For the environment, the information needs are varied, both in subject coverage, scientific and managerial, and geographical scale, local, regional, national and international. In this chapter, three specific topics are discussed in some detail:

● environmental management systems, particularly standards;
● environmental auditing;
● life cycle analysis.

Computer-based information systems

Astute managers will shift their attention from systems to information. Think of the challenge this way: in a competitive world where companies have access to the same data, who will excel at turning data into information and then analysing the information quickly and intelligently enough to generate superior knowledge for purposeful action?

For the majority of managers, information technology and information systems are daunting, tinged with uncertainty and tainted with obfuscating jargon: CPU; MIPS; RAM; DSS; VDU; 4GL; and DBMS. A manager needs to have an understanding of the basic concepts and principles of computer-based information systems, but not an understanding of the technicalities of hardware or computer programming. As managers, we are not interested in technology or systems *per se*. We need useful and relevant information for our decision-making, and, therefore, our concern is how to gather 'data' and how to transform it into 'information'. The categories of data include:

- numbers;
- text;
- image;
- voice.

Although computer data are usually thought of as numbers, decision-making by managers is based primarily on textual data from reports, memoranda, letters and so on. Even though organizations have been using computers for decades, large amounts of (textual) data are not held in a digital, computer readable form. Another important source of information which is not easily made machine readable is through meetings, both formal and informal, and from conversations, both face-to-face and by telephone. Rationality in the use of data, however, cannot be assumed; moreover, data are neither value- nor context-free. The criteria for determining the relevance and quality of data, whether provided by a computer-based information system or by other means, are in terms of their being, or being perceived to be:

- on time;
- appropriate;
- detailed;
- frequent;
- objective;
- comprehensible.

Failure, for instance, to specify the required accuracy for data can lead to unnecessary costs.

Inefficient use of information wastes money. Many examples of inefficiency can be found including:

- information which is collected but not needed;
- information stored long after it is needed;
- useful information which is inaccessible to potential users;

- information disseminated more widely than is necessary;
- inefficient methods used to collect, analyse, store and retrieve information;
- collection of the same basic information by more than one group of people in the same department;
- duplication storage of the same basic information.

(Central Computer and Telecommunications Agency, 1990, page 4.)

While different organizations are at different stages in their investment in and use of this technology platform, the historical perspective can be summarized as:

- improved operational efficiency (since the 1960s);
- increased management effectiveness (since the 1970s);
- enhanced organizational competitiveness (since the 1980s);
- development of total customer service (from the 1990s onwards).

Interestingly, in France, on Minitel, a new, twenty-four hour information service has been created, Ideal Telematique, which provides environmental information.

Computer-based information systems and management decision-making

In the 'classic' work of Robert Anthony (1965), managerial activities were classified into one of three categories, these have been influential in the development of thinking on management of information systems:

- strategic planning (goals, strategies and policies);
- management control (implementation of strategies);
- operational control (efficient and effective performance of individual tasks).

Environmental examples of activities under these categories are given in Table 6.1. The general historical trend is upwards from operational activities towards more strategic applications. Moreover, within an organization, the stage of growth can vary by function (with the accounting/finance function often at the forefront). The incidence of information systems designed specifically for managing our environment are relatively rare, although, where they exist, their focus is usually on operational and management control activities.

Table 6.1 Typical Planning, Control and Operational
Systems

Planning systems	sales (profits/earnings) forecasting operating plans capacity planning new legislation scenarios
Control systems	emissions and energy analyses budgetary control management accounting inventory management
Operational systems	order-entry processing tracking waste documentation transportation scheduling

Using Anthony's classification, a useful framework was developed by
Anthony Gorry and Michael Scott Morton (1989) who differentiated
between the information requirements of management planning and
control activities. Figure 6.1 summarizes the relationships between
different types of decisions and their information characteristics. Planning and strategy decisions, for example scenarios of potential impacts of

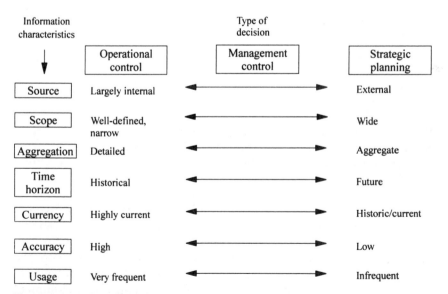

Figure 6.1 Information Requirements by Decision Category
(Source: Gorry and Scott Morton, 1971.)

possible new legislation, are relatively unstructured and require qualitative, as well as quantitative, information. By contrast, many control activities, such as monitoring raw materials, inventories and sales, require simple structured decisions based on quantitative data. Further differences relate to time; a future, rather than a current or historical perspective.

Sources of required environmental data are both internal and external to an organization (see Table 6.2), although it must be stressed that the vast majority of organizations do not co-ordinate or effectively handle, and therefore exploit, their internal data.

Table 6.2   Sources of Environmental Data

Internal	External
Emissions	Legislation
Energy	Global agreements
Waste	market developments
Recycling	Competitive activity
Legislative compliance	Science and technology
Research and development	Pressure group activity
Products	
Expertise	
Skill	
Systems	

(Source : Charter, 1992, page 154.)

Case: United Kingdom's Environmental Information Centre

In 1989, the Environmental Information Centre (EIC) was established to advance the use of computers for the effective handling of large quantities of data in terrestial ecology.

Located at the Monks Wood Experimental Station of the Natural Environment Research Council's Institute of Terrestial Ecology, the EIC is concerned with analysis and interpretation of remotely-sensed imagery, with digital mapping and Geographic Information Systems (GIS) and with the creation and management of large databases in ecology and land evaluation to monitor natural resource management and for land-use planning.

The value of data from earth observation systems can be greatly improved by complementary data from ground survey, from maps and from topographic models. Integrated image analysis and GIS techniques can now be used to combine, to manipulate and to display remotely-sensed imagery and associated digital datasets to produce useful and relevant environmental models.

The practical use of GIS is being demonstrated with the CORINE (Co-ordinated Information on the Environment) programme of the European Commission, which is attempting to harmonize the collection of environmental data in the EC to create an integrated information system on the state of the environment and as a means to guide policy.

The nature of a decision affects the type of information required, and, therefore, the design and specification of computer-based information systems. However, access to good information does not necessarily lead to sound decision-making. The failure to exploit fully information systems has long been recognized as a non-technical issue; the 'information literacy' of managers and the 'information culture' of their organizations are key determinants. In any event, the basic philosophy of information resources management suggests that it is not the quantity of investment that is important, but how well it is managed and used. Moreover, organizations introducing and developing information systems strategies today will be different from those only a few years ago; the technology and the professional knowledge of information resources management have advanced enormously in recent years.

The main reason for the early (and still relevant) investments in information and communications technologies was to improve efficiency, especially in operational areas. More recently, an increasing number of strategic investments have been made, often as 'acts of faith', in the hope that 'IT can be a source of competitive advantage'. For the service sector, John Beaumont and David Walters (1991) have developed a framework for a corporate information strategy comprising four elements:

- *automate* to improve condition and productivity decisions. For example, the use of monitoring equipment to ensure facilities operate at optimal levels.
- *informate* to emphasise the information content of administrative and productive processes within an organization. For example, the use of energy consumption data to provide an input into work scheduling at the level of an individual plant operation.
- *empowerate* to permit action at individual local branches while maintaining head office control. For example, to develop specific local market offers by permitting the use of local management input to introduce a specific local differentiation.

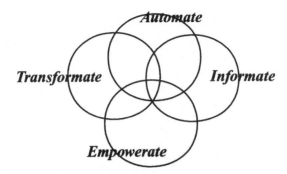

Figure 6.2   The Elements of a Corporate Information Strategy

- *transformate*   to use information and communications technologies to support new methods of doing business and to plan and control new business ventures. They may be applied within the business or externally to make links with customers and suppliers more cost-effective, such as supply and waste chain management.

As alternative strategies are expanded the associated risk can be expected to increase. The purpose of the information strategy is to reduce risk by providing information to relevant management decisions. Within the context of the model proposed, the information strategy is structured to meet both the decision type and information needs of the decision-maker. The four components introduced earlier form the basis of the framework. The corporate information systems strategy is likely to be some combination of each of the four components. They should initially be supportive of each other, overlapping where there is a requirement for interactive support (see Figure 6.2). The four components should be 'balanced' to produce a corporate information strategy. Such a balanced approach would seem unlikely for a particular organization, more likely is the situation where one (or more) of the elements is more prominent.

Case: 'Metadata': Global environmental change

The availability of high quality data is an essential requirement for research into global environmental change. Moreover, researchers must also know what datasets are available, both primary and secondary sources.

To facilitate data exploitation, there is a significant need for information about data, 'metadata'. The majority of the uses of datasets are secondary in the sense that they go beyond the primary use for which they were collected originally.

In the UK, the Inter-Agency Committee on Global Environmental Change has recently established a Global Environmental Change (GEC) data facility with a prime objective to construct and maintain a Master Directory providing a GEC data inventory. It is intended that the metadata, through this GEC Master Directory, should be available both on-line and as floppy disks or CD-ROMs.

This development raises a number of issues regarding details contained in a directory. As a minimum, the level of detail should be by discipline and the natural groupings of data within that discipline. Queries should return all the enquiries containing the combination of keywords specified in the query. Possible fields could contain:

Identification and availability:

TI	proper title (description of content and time/place coverage);
CT	common title (short informal name by which data may also be known);
AU	statement of responsibility for creation (author: person or corporation);
DC	date of creation;
ED	edition statement, which may include a date or revision number;
PU	publisher (statement of responsibility for maintenance/ distribution);
PL	place of publication;
FR	frequency of update (yearly/monthly/regularly/ad hoc);
AV	terms of availability (eligibility for each class of user; price, if any);

Subject and Content:

AB	brief abstract or description of content and origin, a statement; (150 words maximum) and, where appropriate, sub-fields:

	A01	purpose for which data were generated/collected or currently serves;
	A02	population of interest (for example, describe persons or objects);
	A03	period or time coverage;
	A04	geographic coverage;
	A05	extent of aggregation;
	A06	method/instrumentation used to collect or generate the data;
	A07	source of (hard copy) information used in digitizing;

A08   spatial referencing system (for example, Ordnance Survey national grid, unit postcode);
A09   scale;
A10   resolution;
A11   projection scheme;
A12   spatial data structure (statement such as: raster/grid; vector/polygons, lines or points; full topological, raster image or NTF release and structure level);
A13   data quality or processing statement;
SU   suggested keywords or subject terms;

Computing characteristics:

SO   software requirements;
SI   a brief indication of size (such as, 3MB or 2GB);
FO   exchange format (optional);

Access and management:

CO   contact name: statement of responsibility for maintenance;
CA   contact address (postal address, telephone number, electronic mail address);
AK   brief acknowledgement/copyright statement;
DO   citation (title) or reference to relevant documentation.

Information and communications technologies for managing our environment

For effective progress on managing our environment, there is an increasing need for an organization to be able to handle a vast range of internal and external information. As the information requirements grow, and current needs are often not satisfied, new sources of data become important. Martin Houldin at KPMG Management Consulting believes, that in the UK, there are significant developments that are demand drivers for the collection, analysis and presentation of environmental data, including:

● 1990 Environmental Protection Act:
  – Integrated Pollution Control;
  – Duty of Care in Waste Disposal;
● EC Eco-audit;

- EC/UK Eco-labelling;
- EC Integrated Permitting;
- EC Civil Liability;
- British Standards Institution Environmental Management Systems Standards;
- the trend towards more open, external communications of environmental performance in annual reports and other documents;
- increasing interest in environmental performance by different stakeholders;
- the demand for environmental information in merger and acquisition investigations.

Driven by both business and regulatory requirements, in their information systems planning, some organizations are now considering the specification of key environmental performance indicators with the associated data collection/collation.

Case: A Dutch marriage bureau for waste: A value added data service

One organization's waste could be a raw material input for another organization, if only they know about each other.

In the mid-1980s, a company was established in Arnhem, Holland to be a depository of data on available waste, and, therefore, to be able to provide information to organizations that wanted to exchange waste. For an annual subscription, organizations obtain regular information bulletins that advertises both available waste products (called 'residual' products) and required raw materials. Waste human food products beyond their expiry date, for example, can be legally and safely used as animal food. For suppliers of waste, this scheme is often integral to their organization's waste management.

Recent developments include the establishment of an on-line database system for exchange using videotext. For the future, an information and communications technology platform could provide the platform for international waste exchanges.

Environmental management systems

In the 1980s, Total Quality Management (TQM), with the basic aim of ultimately having 'zero-defects', became an important management philosophy and practice leading to many changes in processes and procedures. Linked with a responsibility for the environment, a parallel development is the objective, which is unrealistic for many organizations and industries, of zero negative impacts on the physical environment. As with TQM, especially British Standard 5750, European Quality Standard EN29000 and International Standard Organization (ISO) 9000 series, it is possible to think of quality systems standards for environmental management. Indeed, if general quality systems exist already in an organization, consideration of the environment may be incorporated or could be incorporated relatively easily, rather than viewed separately.

Interestingly, different countries are approaching the specification of standards for environmental management systems. Two basic approaches can be differentiated:

- the development of a new, purposely-designed standard for environmental management systems;
- the adaptation of existing quality standards to incorporate environmental management systems.

By having separate systems, potential conflicts may be minimized. BS7750 is the British Standard on environmental management systems, the first of its kind in the world, which came into effect in 1992. In Holland, extensions to the International Standards Organization 9000 series are being developed, and the Dutch Government has agreed with some of the largest organizations (and the biggest polluters) that environmental management systems will be introduced by 1995. The ISO has established a Strategic Advisory Group on the Environment (SAGE) to advise it on the requirements for and possibilities of the development of environmental standards.

The British Standards Institution has developed a standard for an Environmental Management System, which stresses the need for an organization policy beyond mere compliance (although without emphasizing the requirement to be an integral component of the corporate strategy). The new standard provides a model management system, and aims to meet the growing call from all types or organizations wishing to take a systematic and integrated approach to environmental management with the final aim of improving environmental performance. Figure 6.3 is a schematic diagram of the stages in the implementation of an environmental management system, as viewed by the British Standards Institution. The standard does not specify expected levels of organizational performance, but rather specifies a standardized management system of processes and procedures that is capable of independent assessment and verification.

In more detail, for an organization to be granted BS7750, it must have an environmental management system which is implemented within the

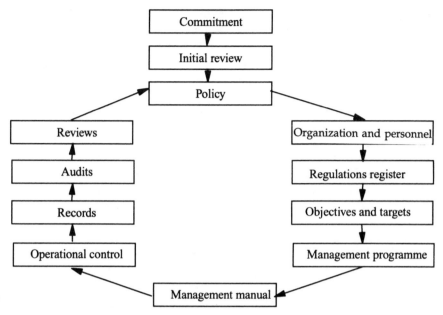

Figure 6.3 Schematic Diagram of the Stages in the Implementation of an Environmental Management System
(Source: British Standard 7750, 1992, page 3.)

broader management systems. An organizational policy statement is a vital ingredient; it must be publicly available and be comprehended throughout the organization. Specific roles and responsibilities have to be given to people, who should possess the appropriate knowledge, skills and experiences; education and training are integral to a commitment for continued improvement. Policy objectives and targets need to be specified in association with the establishment of a register covering environmental effects and legislative requirements. The coverage of the register should not be confined to the organization; upstream environmental considerations of suppliers and downstream considerations of products' uses and disposal. The coverage of the standard is extensive, and, in order to comply, organizations must show that environmental policy is implemented 'from the Board to the shop-floor'. Measurement of environmental performance against targets is necessary, involving record keeping and documentation that would permit possible subsequent verification. A system manual to record future actions and plans is also required.

If successful, BS7750 could offer some competitive advantage for organizations that obtain this environmental quality mark. It could prove to be a powerful marketing tool, and provide the means for industry to demonstrate its commitment to solving environmental issues and for society to judge the environmental performance of organizations offering products, services to consumers or investment opportunities. It could improve an organization's image so that it will also be looked upon more favourably by, for instance, insurers.

As well as national and international initiatives regarding environmental management systems, the European Commission is developing regulations on a voluntary eco-audit, which will require organizations:

- to conduct a review of their environmental performance on a cradle-to-grave basis;
- to publish a statement on its environmental performance;
- to have external verification of the statement.

As planned initially, this scheme will apply to a range of energy, manufacturing and waste industries, but not service industries. Registration can be on a company-wide basis or on a site-by-site basis (see also chapter five).

For success with any environmental management system, standards have to be set against which performance can be measured and audited.

Environmental auditing

Environmental auditing has been in the United States and a few European countries for nearly twenty years. In the United Kingdom, it is a relatively new activity, especially outside of the chemical and oil industries. The Confederation of British Industry defines environmental auditing as,

> ... the systematic examination of the interaction between any business operation and its surroundings. This includes all emissions to air, land and water; legal constraints; the effects on the neighbouring community, landscape and ecology; and the public's perception of the operating company in the local area.

The term, environmental auditing, is defined and used in a variety of ways, and, in practice, environmental audits vary enormously in scope and detail. An environmental audit is a management tool for the assessment of an organization's overall environmental performance. It is not an Environmental Impact Assessment (EIA), which considers the potential impact of a proposed project. Not surprisingly, environmental audits are often thought to have similarities with traditional financial audits. While there are suggestions, including EC proposals, that would make them much more similar, the fundamental characteristics of a financial audit, an annual, statutory requirement to publish with independent attestation of the statements, are not found today in environmental auditing, which remains voluntary and has no regular timetable. Moreover, there are no agreed standards for environmental auditing in Britain, and they would be necessary if attestation became part of environmental auditing.

As indicated by Table 6.3 categorization of inputs, processes and outputs, an environmental audit is an intensive information processing activity. Environmental auditing should not be viewed as an isolated activity, rather it should be integral to both the strategic and operational management aspects of an organization. Indeed, it is often important to collect data on a broader range of issues as summarized in Figure 6.4.

Table 6.3 Environmental Performance Review

Inputs	Processes	Outputs
Renewable resources	Waste minimization	Product impacts during use
Sustainable extraction	Pollution control	Energy consumption
Secondary effects (such as deforestation)	Energy use	Packaging
Energy sources and use	Waste management	Recyclability
Transport	Building and plant maintenance	Disposal
Suppliers	Transport	Potential resource recovery

(Source: KPMG Peat Marwick McLintock.)

Whether an environmental audit is completed by internal people or by external consultants is really not an issue, as long as the team possesses the required and varied specialist skills and experiences; an environmental audit is undertaken for internal management. Although organizations do undertake comprehensive environmental audits, specific assessments are undertaken for particular circumstances, including:

- acquisition audits that can examine possible liabilities of proposed mergers and acquisitions;
- compliance audits that are focused on existing and expected legislation;

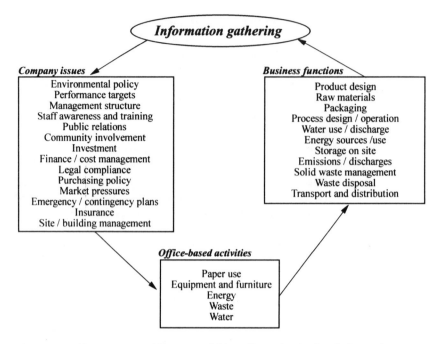

Figure 6.4   Environmental Issues and Your Organization's   Information Gathering
(Source: Coopers & Lybrand Deloitte, 1991, page 11.)

● supplier audits that cover the environmental performance of suppliers.

In summary, any environmental audit is driven by the specified objectives, and attention is given to existing processes, procedures and practices, including their risks, with recommendations, as appropriate, for changes. An environmental audit should be viewed as a first step!

The influence and incidence of environmental auditing varies by county and by industry and organization. It is likely to increase in significance in the forthcoming years, although the costs of completing the exercise may be prohibitive for many organizations. The decision to act with regard to the environment is important, but many commentators, particularly pressure groups, argue that without a commitment to specify objectives and then to measure, monitor and publish, it is unsatisfactory and hollow.

Case: Your business and the environment: A D-I-Y review for companies

The challenge for individual businesses is to understand what environmental performance means for them and to be able to develop appropriate action plans. Working with Business in the Environment, Coopers & Lybrand Deloitte (1991) produced an excellent, practical workbook, particularly for small and medium size businesses, as a detailed guide to conducting an environmental review of a business.

The guide covers the following tasks:

- look at:
  - issues: what is at stake?
  - regulations: what is required?
  - standards: what could you achieve?
  - examples: what are others doing?
- develop a framework environmental policy;
- carry out a total assessment of the current environmental position and performance of your company. Identify the strengths and opportunities, as well as the risks and threats.

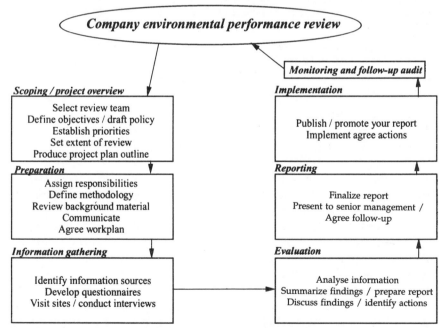

Figure 6.5 Environmental Performance Review Procedure
(Source: Coopers & Lybrand Deloitte, 1991, page 10.)

Life cycle analysis

An overview of life cycle analysis, particularly its comprehensive perspective, is presented in chapter five. However, the implementation of the underlying principles are data rich, and it is appropriate to examine some practical issues in this chapter. At the outset, it must be emphasized that no single set of tools exist to analyse environmental impacts at each stage of a product's life. Moreover, it is stressed that life cycle analysis should not be seen as an end in itself; instead, it should provide invaluable inputs for more informed management and consumer decision-making. For example, life cycle analysis is being adopted as means of implementing ecolabelling standards to compare the environmental impacts of competing products.

With life cycle analysis developing in the 1980s, it is increasingly viewed as more than a simple, integrated inventory of resource and waste inputs and outputs, a kind of balance sheet. Economic, as well as local cultural and political, dimensions are important characterisitcs of life cycle analysis, with public policy significance as well as being of business importance.

Case: 3M's product responsibility guidelines

A kind of life cycle analysis has been developed by 3M to examine its planned products.

Their so-called visualization model has been introduced to provide a consistent foundation to explore the environmental impacts of new products. While being refined with experience, the key characteristics of the visualization model include:

● *product concept:*
  - define customer performance requirements;
  - anticipate regulatory/safety/public requirements.
● *product design:*
  - component selection:
    * toxicity; use reduction;
    * renewable resources;
    * energy intensity;
    * safety;
  - packaging:
    * minimal use;
    * renewable; recycled;
    * recyclable;
    * toxicity; disposability;
  - product risk assessment.

- *process:*
  - waste and emission reduction;
  - safety/health;
  - efficient RM and energy use.
- *use:*
  - safe handling recommendations;
  - identify/address misuse;
  - waste and emission reduction;
  - energy requirements;
- *disposition:*
  - reuse, recycle;
  - degradability;
  - adverse environmental impact;
  - disposal recommendations.

This corporate development is significant, because 3M produces thousands of products and over a quarter of its total revenues comes from products introduced in the last five years.

The biggest task in relation to life cycle analyses is to reach common standards and assumptions behind the measurements; the requirement is for an unbiased methodology and comprehensive databases, which make it possible to objectively compare companies and their products. Sound progress has been made to date, but there should not be any methodological complacency. In some circumstances, a complete life cycle analysis is either infeasible or uneconomic. Significant practical problems relate to the necessary level of detail that is required in any particular analysis, and to the definition of a problem's boundaries. For example, it is frequently questioned whether a comparison between cloth and disposable nappies (diapers) should include an evaluation of the environmental affects of manufacturing and using washing machines to clean cloth nappies or of timber operations to produce the paper for disposable nappies.

Tools that consider the environmental impacts of products and process techniques and monitor and audit environmental performance are vital if a comprehensive and co-ordinated environmental policy is to be incorporated fully into an organization's corporate strategy. Migros, a well-known Swiss retailer with a strong environmental tradition, for example, markets software, called Eco Base 1, which automates the use of life cycle data for assessments of alternative packaging designs.

Given the increasing interest across an industry's supply and waste chain, it is not surprising that many organizations are now using questionnaires and means of collecting data to obtain information from their suppliers and their suppliers' suppliers. This is a non-trivial task for a life cycle analysis; for instance, IBM is already collecting data at fifteen levels away from its direct suppliers! The world's largest tissue manufacturer, Scott, completed recently a life cycle analysis of its products, which indicated that the most beneficial environmental impact could be made in wood pulp manufacture. As a consequence, Scott asked their suppliers about their environmental performance: ten per cent of its worst suppliers have been discarded; and the best suppliers are being given preferential treatment.

Concluding comments

The quality of available information and the quality of its management will be important determinants of the success of managing our environment. Environmental information is increasingly becoming a strategic resource, and benefits and power will be derived from how they are used, rather than merely having access to them.

> Today, the big thing is not how to use the computer, or even how to organize information, but how to organize one's own work in the information-based organization. This idea is something that we are just beginning to nibble on.
>
> (Drucker, 1990, page 73.)

While the facilitating technologies are obviously important for information, it is becoming increasingly clear that successful exploitation of this strategic resource is dependent on the attitudes, knowledge and skills of the available people.

More information and greater openness and democracy are vital for managing our environment. There is a requirement to put into practice fundamental rights and freedoms, such as the EC's recent Directive on the Freedom of Access to Environmental Information (although this can exclude data held by the European Commission itself!).

Review questions

1. For your current organization, define and justify your environmental information requirements.
2. Describe what you understand by an environmental audit, and indicate how it should be able to assist senior management.
3. Consider the linkages between Environmental Management Systems and Total Quality Management.

Study questions

1. For an organization that you know, specify its environmental information needs and the associated information sources.
2. With regard to the global management of our planet, examine the requirements for international data exchange and mechanisms to promote effective information resources management.
3. For a product of your choice, outline how you would complete a life cycle analysis.

Further reading

Beaumont, John R. and Sutherland, Ewan (1992) *Information Resources Management*, Butterworth-Heinemann, Oxford.

Information Technology is too important to be left as the responsibility of the IT specialists. In considering management in our knowledge-based economy and society, John Beaumont and Ewan Sutherland provide a non-technical discussion for managers. The coverage is founded on the authors' belief that an information strategy must be an integral component of the corporate strategy, and that information flows are the 'blood' of organizations and should be able to assist management decision-making. No manager can avoid the implications and impact of information and communications technologies both on management processes and ways of doing business.

Emery, James C. (1987) *Management Information Systems*, Oxford University Press, New York.

A clear, balanced introduction for managers wishing to understand and exploit the potential business benefits of computer-based information systems. It is part of the Wharton Executive Library and satisfies the objectives of the series, that each volume is:

- up to date – reflects the latest and best research;
- authoritative – authors are experts in their fields;
- brief – can be read reasonably quickly;
- non-technical – avoids unnecessary jargon and methodology;
- practical – includes many examples and applications of the concepts discussed;
- compact – can be carried easily in briefcases on business travel.

Mintzberg, Henry (1989) *Mintzberg on Management*, Free Press, New York.

An excellent, clear discussion of 'inside our strange world of organizations'. Many challenges are made about the conventional wisdom of management, based on Henry Mintzberg's experiences of 'bottom-up management'. The usefulness and relevance of information is placed in its organizational context of strategy, structure and power. (All MBA students should read Chapter 5, 'Training Managers, not MBAs'.)

Synnott, William R. (1987) *The Information Weapon*, John Wiley and Sons, New York.

A clear book for business managers, with a wide range of different case studies, which places IT/IS at the strategic level. In total, 63 information weapon strategies are proposed!

7 Corporate culture, identity and human resources management

Looking to the future of business, the ability of a company to adapt quickly and harmoniously to an environment that is going through a process of radical change affecting basic beliefs and assumptions – a paradigm change – depends more than ever upon the success of all levels of management in creating and sustaining excellence in innovation. In so far as we try to come to terms with those new beliefs and assumptions, both vision and values will need to reflect them.

... The change in direction of company activities can only come about smoothly if a thoughtful effort is made to change the vision and value systems so that they confirm to the conditions for sustainable development that affect any particular business. The choices made will explicitly recognise the priority of meeting the needs of people, whether they be customers, employees or other contributors to the enterprise. The value systems will be people-centred, and the vision will foster a co-operative excellence in innovation that is sensitive to the necessity of conserving scarce resources, while minimising waste and environmental or ecological damage.

(John Davis, 1991, pages 55–56.)

Introduction

In this chapter, an understanding of the cultural, organizational and social issues related to the successful management of our environment is provided. These issues are of growing importance, with much more than a purely scientific significance; they need to be set in the context of organizational policies for Human Resources Management (HRM). For managers, environmental performance can only be achieved in business terms. Ultimately, the success depends on the attitudes, skills, knowledge and experiences of the people involved in the development and implementation of environment policies. Like other assets, human managerial expertise is essential to maximize the exploitation of the business opportunities from managing our environment; it is the people in an organization who identify the problems, determine how to solve them and implement necessary actions. Moreover, it is the culture of an organization and hence its people that will directly affect the responsibility they believe their organization should have to better manage our environment. Environmental 'excellence' will only occur if there is real commitment and action by everyone in the organization.

In the context of managing our environment, the relationship between business strategy and the organization, its human resources and its culture is important, because people determine directly the success or otherwise of any strategy. An organization interacts with the business strategy in the sense that the organization must deliver the results of the strategy and the organization also evolves the strategy. In that respect, it is important to foster an organizational culture which is supportive of the organization's future activities, including a commitment to the environment. Without a match between strategy and culture, business success is unlikely. An organization also interacts with the physical environment, through its use of resources and production of waste and pollution; the manner of this interaction is dependent on the specific organization's activities, products and services and its environmental culture. The logical connection lies in a future of greater concern for the environment and in the vision of where the organization is going and how it is to get there.

This chapter includes a broad ranging discussion of organizational culture, identity and human resources management. The foundation is provided by an examination of corporate culture, the patterns of assumptions which guide the people in an organization and their behaviour. More specifically, for successful and better environmental management, it is essential to explore how an organization's culture can be changed, particularly through:

- *Human resources management:*
  – the mangement decisions which affect relationships between an organization and its people.
- *Leadership:*
  – how an organization can be modified by dynamic leadership/top management action.
- *Structural implementation through organizational change:*
  – how our knowledge of an organization and of environmental pressures can be used to increase the chances of success in the implementation of management for our environment, supporting the business strategy and corporate performance.

In addition, and linked to the discussion in chapter three, there is a consideration of more general issues of corporate social responsibility, including stakeholders' needs and wants, and corporate governance. Attention is also given to the specific need of education for our environment, as an integral component of people and organizational development. Failure with regard to managing our environment, either to act or to achieve successful implementation, is likely, increasingly, to significantly impair the corporate performance of an organization. In extreme cases, it could make it difficult for an organization to recover, leading to a downward spiral in performance. However, this is not to suggest a scientific imperative. The real dangers lie ultimately in deeper non-scientific issues, concerning organizational, cultural and social matters.

Corporate culture

A number of ideas from the social sciences have been brought into organization theory and a few have even made the transition into management practice. Of these ideas, organizational culture is one of the most useful, though it remains difficult to grasp and define. Simply stated, the culture should not be seen as something an organization has, rather culture is what the organization is. Organizational culture is seen as a combination of deep-set beliefs, a shared vision, about the way work should be organized, the way authority should be exercised, and the ways in which people should be, and are, rewarded.

> A pattern of basic assumptions – invented, discovered by a given group as it learns to cope with its problems of external adaptation and internal integration – that has worked well enough to be considered valid and, therefore, to be taught to new members as the correct way to perceive, think and feel in relation to those problems.
>
> (Schein, 1985, page 9.)

Organizational culture is a way of looking at organizations in terms distinct from their structures, processes and procedures. The use of organizational culture allows us to go beyond a hierarchical or a bureaucratic view, it allows us to consider social interactions in management processes. The culture is how values and rituals are transmitted to members of the organization through a 'collective mental programming', for example, how hard and how long we should work: should we stay late, take work home and come into the office at weekends.

> Consider IBM breakfast, for instance. You'll never see IBM salespeople along with the hordes of others congregating at Howard Johnson's every morning, because IBMers are encouraged to see their time as too valuable to waste in

a roadside diner. When IBMers want coffee they will share it with a client or colleague. The point is for IBMers to begin the day not discussing baseball or the price of steak but to get a head start by focusing on the company, the industry, and their habits as professionals.

(Deal and Kennedy, 1988, page 60.)

The importance of corporate culture in the management process has been highlighted since the beginning of the 1980s. In attempts to better understand the complex concept, several definitions have been put forward. Despite variations, most definitions agree that organizational culture should be seen at several levels:

● *artifacts, symbols and creations:*
  – these include the 'visible'/tangible objects such as buildings;
● *values and attitudes:*
  – these are testable in both physical and social environments and include art and patterns of behaviour;
● *basic assumptions and core beliefs:*
  – these concern the organization's 'world view' and its relationships with the business environment, people and so on.

Although certain aspects of the organizational culture can be seen, the 'hardware', they are often difficult to decipher, while the basic assumptions, the 'mental software', are taken for granted and cannot easily be tested. At a superficial level, simple considerations such as standards of dress are often the most noticeable. For example, academic staff in universities can dress very much as they like, idiosyncrasy is almost encouraged, whereas in companies such as IBM, there is a dress code; relatively strict rules have long been in operation requiring male staff to wear conservative suits, white shirts and striped ties. Similarly, symbols such as buildings and the appearance of offices can be very revealing. In the UK Civil Service, offices have a degree of uniformity which is almost unknown in the private sector. Office furniture and equipment is purchased nationally and distributed as needed. Each grade of officer knows what he or she is entitled to, in terms of carpet or linoleum, number of chairs and so on. Thus, on entering the office of civil servant whether in the Treasury in Whitehall or in the Scottish Office in Edinburgh, you find offices which are almost identical, down to the same pattern of chair covers. It is possible to tell the grade of the official you are meeting by a quick scan of the office.

The most impressive and potent artifact in the whole of the UK Government is the portrait which sits in the office of the Minister of State in the Foreign and Commonwealth Office; it is of George III, the monarch who lost the United States of America. While such patterns are interesting, the real question is what they tell us and what we might do as a result of such knowledge. Japanese companies, such as Toshiba, require all staff, including managers, to wear uniform jackets so that they cannot readily be distinguished, making the single status of staff evident to all. This acts to break the social and organizational distance between manager and worker so common in American and European organizations which so often leads to aloofness and insensitivity.

Stories, jokes and, at a deeper level, myths are important ingredients of the culture of organizations, transforming events in an organization's history into a valued heritage and a source of shared meanings. They also turn bureaucracies into more human places, offering a break from the daily routine and permitting the display of emotions such as amusement, pride, bitterness, relief or worry. Myths and jokes also provide a way of expressing hard truths which might otherwise go unheard. For this reason, they can open windows into organizational life and reveal unnoticed aspects. One of the key ways in which an organizational culture is passed on is through stories, which often identify heroes and anti-heroes within the organizations.

The significance of organizational culture in the management of transformation must not be underestimated. One of the many ways in which it has been suggested that managers might improve their organizations is through changes in the organizational culture. Populist authors, such as Tom Peters, have suggested that the organizational culture can be engineered, in the belief that particular cultures are better suited to particular fields of activity. Indeed, their 'excellent'/successful companies of the 1980s all had a strong culture. One of the most difficult things to achieve in an organization is innovation, which Peters and Waterman (1982, page 120) tried to show to a client by visiting 3M to see how they went about problem solving,

> Our friends at 3M were tolerant of the excursion, and we observed all sorts of strange goings-on. There were a score or more casual meetings in progress with salespeople, marketing people, manufacturing people, engineering people, R&D people – even accounting people – sitting around, chattering about new-product problems. We happened in on a session where a 3M customer had come to talk informally with about fifteen people from four divisions on how better to serve his company. None of it seemed rehearsed. We didn't see a single structured presentation. It went on all day – people meeting in a seemingly random way.

The ideas of quality and service, which have so dominated management thinking in recent years, have important implications for the organizational culture and place significant demands on it. For instance, looking to the transport sector in the UK, one sees the need for change, not only in investment but also in changing the attitudes of staff who have often lacked courtesy and have frequently seemed to view passengers as at best a nuisance. To change such deep-seated attitudes requires a shift in organizational culture – a creation of new myths. While British Airways (BA) is now less frequently known as 'Bloody Awful', its counterpart, British Rail, is still trying to get there!

Corporate culture is very difficult to change and the time scale involved can therefore be significant. By comparison it is easy to change a business strategy and relatively easy, if expensive, to change the technology platform. On a superficial level, one can issue new dress codes and, if necessary, appear in jeans and T-shirt or even in white tie and tails. The real challenge is how to change values and beliefs. However, culture is not static; it should be seen as an adaptable/tangible learning process. On the other hand, a strong and inflexible culture can equally act as a barrier to further development, a fact which in part might help explain IBM's recent decline in performance.

For the average European manager, the books on organizational culture may have a lot to say about organizations, but rather less about what they perceive to be 'culture'. Many of the examples from American organizations appear to be ghastly excesses of 'hoopla', causing the European reader to cringe. Whether because Europeans see themselves as more refined or because they are more conservative, it is inconceivable to imagine typical French, German or Scottish managers behaving in ways which their North American counterparts would regard as quite acceptable. Nonetheless, after toning down, the American message is valid. Culture is changed by group activities, by coming together to do things and to celebrate things.

The link between corporate culture and national culture is important, and, in this discussion, it is sufficient to state that many people believe that national culture is an important driving force of corporate culture. Someone has only to have worked in a multinational's offices in London, New York and Madrid to appreciate that there are significant differences in these offices' culture, which, in part, can be attributed directly to differences between national cultures. For the environment, it is interesting to note Adrian Webb's (1991, page 1) comments,

> Two previous reports on environmental policies in France and Germany drew attention at an early stage to the deeply instinctive understanding of what is meant by the environment, characteristic of each country. This report makes no apology for doing the same . . . . It is also important to remember that attitudes are not uniform throughout the whole UK. Both Scotland and Wales have distinct cultural traditions, and there are significant variations within England itself, particularly between north and south.

For instance, different impressions, picturesque gardens, wild moorlands, highlands, and so on, are conjured up by different people, depending on their backgrounds.

The key is to shape the organizational culture in such a way as to empower people to use the new vision. In the first instance, this is done by asking questions which empower 'what should we do?' and 'how can we help you?' The celebration of success is a major opportunity to reward people who have acted in a way which fits with the new culture. The important factor is to generate concrete actions to support the vision. One important use of the organizational culture is to overcome the failings of the structures and procedures of the organization, which also, unfortunately, often drive or at least constrain strategy. That is, to have a culture which tells people not to adhere to the rules and procedures so rigidly that it is going to hurt or upset the customer. This is in complete contrast to the bureaucratic approach which is to embed the rules in a manual and not recognize the ability of the individual to override those rules.

Case: Générale de Service Informatique

Scenario: Générale de Service Informatique has achieved considerable growth while retaining an unusual organizational style which leaves considerable responsibilities with individual managers and business units.

Générale de Service Informatique (GSI) was launched in 1971 initially to provide a computer bureau service. It expanded rapidly through the enthusiasm of a group of individualists who refused to conform to the traditional norms of large organizations.

The main sectors of GSI's activity are: management, motor trade, transport and tourism, marketing information, banking, facilities management, advanced technology, payroll and human resources management. GSI uses very simple organizational structures and procedures with autonomous operational units. Despite many problems, GSI has been able to succeed with its non-bureaucratic system. Perhaps the most important feature of GSI is that its organizational culture is sufficiently strong to correct the adverse effects of the rules and the structure (or the lack of them).

In a traditional hierarchical organization, decisions must be referred upwards. This means that a manager cannot always make decisions for his staff which, in effect, emasculates the manager. GSI devolves power to managers to avoid this problem.

In terms of organizational structure, GSI has no personnel department; responsibility for personnel lies with managers. In terms of organizational culture, individuals at GSI are encouraged to value their colleagues, for example, to listen where once they would have talked. The overall aim of GSI's HRM policies is to create a set of conditions which are favourable to individuals, which allows them to develop and to use their abilities to the full. People work as teams and develop habits of thinking and acting in common. The role of a manager is to:

- listen;
- inform;
- delegate;
- facilitate.

This is reflected in the criteria used for assessing managers:

- Have you obtained the best from your colleagues?
- How have you conducted yourself vis-à-vis your peers?
- How have you helped your boss?

(Source: Crozier, 1990.)

A strong organizational culture for environmentally 'excellent' organizations in the future will be as important as it has been for successful companies in the past. What is needed is management change as managers must act as 'opinion leaders'. It must be identified:

What organization do we think we are?
What are we actually?
What kind of organization would we like to be?

The question is not if managers are able to change the culture, rather it is most often a question of their willingness to do it.

The organizational culture contains values and basic assumptions, information about what the organization is, about its mission and its *raison d'être*. These values and assumptions indicate how the organization operates with respect to other organizations in the business environment in terms which can be grouped together as business ethics and environmental care. The attitudes towards its staff, its shareholders, its customers, its suppliers as well as towards the community in which it operates. Anthony O'Reilly of Heinz, opened his Chairman's Statement, 'Year in View', in the 1991 Annual Report by stating,

The last year has been one of change and challenge. In trading terms, the recession has caused reduced consumer demand but the company has contrived to perform strongly.

On the legislative and regulatory front there has been considerable activity both in the UK and within Europe. A new Food Safety Act has been enacted, an Environmental Protection Act has been added to the statute books a wide ranging White Paper on the environment published. European legislation awaits in the form of Eco-labelling and packaging directives, food labelling proposals and discussion on formalizing environmental audits.

Overall environmental issues have remained high on the agenda, and the public's consciousness of environmental matters has matured with a more discerning attitude being taken by some consumers.

In May 1990 an Environmental Audit was conducted across the Company. As a result, an Environmental Team was set up with senior representatives from all functions with the brief to execute a detailed Company-wide environment strategy for the company.

Human resources management

In recent years, human resources management (HRM) has become one of the most important of the 'topical' areas in business, along with 'total quality management', 'excellence' and so on. This development has been fuelled, in part, by business and management books of the kind sold in airport bookstalls. Nonetheless, there are real business issues which have pushed HRM to strategic importance:

- increasing global competition;
- increasing size and complexity of organizations;
- growing demands for career and lifestyle satisfaction;
- changes in workforce profile (age, nationality, sex and skills);
- more legislation.

HRM is now seen to be a vital component of an organization's strategy. Changes in the business environment and in the way in which managers perceive their organizations have raised HRM to a level of importance which was never achieved by either personnel management or industrial relations. This section includes a discussion of:

- *Network organizations* how the new organizational structures, which go beyond conventional organizational boundaries, affect HRM.
- *Intellectual capital* how the knowledge within an organization can be used to create differential advantage in the marketplace;
- *Education and training* how the abilities of the people in an organization can be developed.

One of the better definitions of human resources management is provided by Michael Beer and his colleagues (1984, page 1),

> Human resource management involves all management decisions and actions that affect the nature of the relationship between the organization and employees – its human resources.

HRM is much, much more than personnel management and industrial relations. Personnel management can conjure up images of third rate managers sent somewhere where they 'can do no harm', checking superannuation details or completing government statistical returns, while industrial relations can suggest the bizarre rigmarole of collective bargaining which precede acquiescing to the petty tyrannies of trades unions and the loss of yet more international competitiveness. It is interesting to note complaints from students, in both the UK and the US, that industrial relations is taught as case law, that is, what you are legally allowed and forbidden to do, while what should be done is to identify best practice. Michael Beer and his colleagues (1984) subdivide HRM into:

- *influence of people*   employees are 'stakeholders' with interests in the organization which are important in considering how an organization functions; these range from salary and other financial rewards to dignity, satisfaction and status;
- *human resource flow*   the 'flow' of people in, through and out of an organization must be managed to produce the right mix of knowledge and skills, if the organization is to meet its strategic objectives;
- *reward systems*   rewards are important to attract, motivate and retain staff. Questions which arise include remuneration for individuals or groups, for example, should there be profit sharing and how should cash rewards be used: status, goods and services?
- *work systems*   the definition and design of work.

If these four areas of HRM policy are effective, then the benefits should be found in terms of:

- *commitment*   people should share and be committed to their colleagues and to the clearly stated and well understood goals of the organization;
- *competence*   people should have the levels of attitudes, skills and knowledge necessary to carry out their specific tasks and more generally to carry out the mission of the organization;
- *cost-effectiveness*   the overall cost of the people should be close to the lowest possible cost commensurate with achieving the corporate mission and without causing unnecessary problems.

Case: Toshiba in the United Kingdom

Scenario: in 1981, Toshiba took over a failing television factory in the UK and made it into a success through non-traditional management techniques.

On 28 October 1986, Toshiba Consumer Products celebrated the production of its one millionth television set from its plant at Plymouth in South-West England. The plant had been opened in the late 1940's by Bush to manufacture radios. In 1972, it passed to the Rank Organization, a UK-based conglomerate which is a financial partner in Rank Xerox. A joint venture with Toshiba, the Japanese electronics giant, was established in 1978. The failure of the joint venture in early 1981 allowed the Japanese to take over the plant. The blame lay largely with Rank, which had not adopted any Japanese or American management ideas and had consequently failed to exploit the technology made available from Toshiba. While Rank wanted to produce television sets which were easy to maintain, Toshiba had aimed to produce television sets which did not require any maintenance at all.

The new Japanese operation was markedly different from the joint venture:

● four factories cut to one;
● 2,600 'workers' cut to 300 'members';
● seven trades unions cut to one.
● sixty-two models cut to eight;
● circuit boards used in a TV, cut from twelve to three;
● manufacturing time cut by 40%;
● output increased from 300 sets per day to 2,000;
● video cassette recorders and microwave ovens added to production.

These changes were possible as a result of the crisis caused by the closure of the original operation, they were certainly not incremental. The 'new' factory was consciously different from the old one. However, the management practices were not strictly Japanese and they were certainly not magical:

● select and train 'members' for flexible working;
● focus production on television sets;
● develop new relationships with suppliers emphasizing quality, pricing and timing of delivery;
● introduce single status employment contracts;

- introduce single union agreement with 'pendulum arbitration';
- create a Company Advisory Board (COAB) to discuss all aspects of company performance with 'members'.

The plant operated without job descriptions and 'personnel' functions were the responsibility of line management. The single status of 'members' meant that:

- all 'members' wore similar blue company 'coats' with Toshiba badges;
- there was a single staff restaurant;
- there was a single open-plan office for all staff, including the managing director;
- all staff were paid monthly.

From the start, Toshiba Consumer Products operated as part of Toshiba's world-wide operations, integrating into its sales and marketing networks.

(Source: Trevor, 1988.)

Network organizations

In our discussion of competition and business strategy in chapter two we have shown how some organizations now operate through collaboration across industry supply and waste chains. No longer do these organizations rely on simple contracts for purchases and sales, they have developed complex partnerships, collaborating in joint projects. In such networks, interrelationships between individuals in a single organization or in different organizations become much more significant than the formal structure of the organization which officially govern the links. Jack Rockart and Jim Short (Scott Morton, 1991, pages 189–219) identify the key attributes of a successful networked organization as being the sharing of:

- goals;
- experience;
- decision-making;
- responsibility;
- accountability;
- trust;
- timing;
- work;

- recognition;
- rewards.

This new multiplicity of networks achieves business goals in non-traditional ways, which presents challenges in terms of managing the collaboration and interdependence of the individuals, business units and separate companies to achieve the overall corporate strategic objectives. A critical capability is provided by the technology platform which can allow the co-ordination of the many individuals and different business networks. Networked organizations need to find new measures of performance which take into account the diffuse and co-operative nature of business, addressing the particular problems for assessing the performance of individuals. These organizational issues are obviously integral to successful cradle-to-grave management and collaboration across inter-organizational boundaries for supply and waste chain management.

Intellectual capital

As organizations struggle to compete they have been forced to realize that the most important way in which they can increase their competitiveness is through the effective utilization of the people they employ. This can be termed 'intellectual capital', which Hugh Macdonald, formerly of International Computers Limited, defines as,

> . . .knowledge that exists in an organization that can be used to create differential advantage.

Organizations which use their 'intellectual capital' more effectively than others are able to realize competitive advantage through their ability to achieve differentiation not once, but repeatedly and regularly. Competitive edge comes from applying the knowledge and skills of individuals in the organization in order to achieve intangibles such as being perceived as being 'talented', in terms of:

- research;
- development;
- design;
- tailoring.

The concepts of managing intellectual capital and its practice are still in their infancy. For managing our environment, given the relative lack of experienced people, this issue is of fundamental importance. While the need for environmental specialists, scientists, lawyers and so on, is obvious, the real need is for an organization's people to all have much greater environmental understanding. The increasing empowerment of the line has direct implications for new skill requirements, if implementation is to be successful. For instance, in terms of skills shortages, in Britain, the Association of Metropolitan Authorities believes that policing new environmental legislation is likely to produce new and extensive pressures on local authorities, especially those in London and the South-East. Legislation in Germany requires major polluting companies to appoint an Environmental Protection Officer, who is responsible for pollution control and management. At present, however, only the larger organizations seem to have done something about giving someone full-time responsibility for managing our environment.

Clearly, it is necessary to reward and retain those who possess the individual building blocks of intellectual capital if the organization is to utilize their abilities. Interestingly, there is increasing evidence that many people are unwilling to work for organizations with poor environmental images or practices. Moreover, intellectual capital depreciates and must be renewed, not only through new people entering the organization, but also through those already there learning through new activities. Intellectual capital must be identified and fully exploited. A key consideration is the ability to generate and to carry new ideas through into practice.

Education and training

If an organization is to survive and prosper, it must ensure that its people have the attitudes, knowledge and skills necessary to carry out its corporate mission. It must also be possible to develop a new mission and to carry that out. The development of people's attitudes, knowledge and skills in the organization is essential to allow it to exploit business opportunities. To instil an environmental consciousness in an organization, staff training and development must take place at all levels, because all the people must understand not only the importance of environmental performance, but also their organization's environmental commitment.

General debates on education have been undertaken in most European countries and in the US and Japan. Each country believes itself disadvantaged and is striving to improve its national system, usually by copying features of another. In Europe, there is also a slow drift towards a single system as mobility of students and managers increases.

However well educated people are on their recruitment to an organization, it is vital for organizations to develop their talents. Here, Japanese companies are recognized as the leaders, with extensive programmes for staff development. Looking at the in-company training of organizations in the UK, Coopers and Lybrand concluded,

Most companies agreed that Britain did under-train compared with its main overseas competitors, but also thought that the amount of training they themselves undertook was about right. This lack of concern might be regarded as reflecting confidence; we think it would be more realistic to regard it as reflecting complacency.

This complacency was reinforced by a widespread ignorance among top management of how their company's performance in training compared with that of their competitors - even those in the U.K., let alone overseas.

(Coopers and Lybrand, 1985, page 4.)

One of the main reasons why organizations will not train their people is because they are afraid that other organizations will poach them. So in order to acquire the necessary knowledge and skills, they poach people from other organizations, creating a vicious circle. A more positive way of thinking would be to create an attractive and stimulating working environment to ensure an organization retained its people with their skills and knowledge.

Case: Environmental training: What is needed and how should it be provided?

New environmental legislation and pressures from various stake-holders are placing increasing requirements on businesses to transform the way that they manage themselves. These forces are generating a need for new skills in a range of areas, including:

- integrated pollution control;
- waste management;
- contaminated land assessment;
- environmental auditing;
- life cycle analysis;
- packaging;
- environmental legislation.

To help satisfy these needs, organizations can:

- recruit people with appropriate skills;
- train their existing people;
- employ consultants.

Aware of the demand from business for greater environmental knowledge and skills, a recent report on *Environmental Responsibility*, which was commissioned by the Department for Education and the

Welsh Office (1993), explored the agenda for further and higher education institutions.

It was argued that these institutions should have an indispensable role by offering:

- specialist courses leading to specifically environmental qualifications;
- 'updating' courses for those already in the workforce;
- environmental education for all students, whatever their specialist subjects of study (so-called 'cross-curricular greening').

Leadership

It is no longer plausible to believe that business success can be achieved through the traditional hierarchy of authority, rules and procedures for co-ordination and the division of labour; the division of tasks into small parts and their specification in manuals (noting that at one time the UK Post Office manual required over three metres of shelf space). The tasks organizations set cannot be specified simply in terms of actions, but require that we adhere to sets of values held in common with each other and with the organization. Customer service, quality, process innovation and so on, cannot be achieved by direction alone, they require acceptance of certain common values. Experience of management shows that people must be led.

> To manage is to lead, and to lead others requires that one enlists the emotions of others to share a vision as their own.
> (Henry Boettinger, formerly Director of Corporate Planning, AT&T.)

> People cannot be managed. Inventories can be managed, but people must be led.
> (H. Ross Perot, Founder, Electronic Data Systems and Presidential candidate in the United States of America.)

> Making it happen means involving the hearts and minds of those who have to execute and deliver. It cannot be said often enough that these are not the people at the top of the organization, but those at the bottom.
> (Sir John Harvey-Jones, formerly Chairman, Imperial Chemical Industries.)

Nowadays, we are all leaders, whether we lead from 'below' or from 'above'. We must identify and develop change in our own organizations and must help and encourage our colleagues, clients, suppliers and partners to do the same. We must encourage those who are practising change and, most importantly, we must welcome change ourselves. There is nothing worse than for people to know that they can sit out your current fad.

The manager as consultant

The role of the manager as described in this book is dramatically different from the traditional view. He or she is now called upon to play roles such as:

- coach;
- counsellor;
- leader;
- mentor;
- problem finder;
- problem solver.

The old-fashioned roles of the manager were as an expert and commander, telling people what to do and providing solutions. Today, it is unlikely that this role will succeed or even be possible. By comparison, the traditional role of the consultant has been to appear, ask a few questions then write a report with definitive conclusions. It is increasingly likely that the consultant will be required to stay around and implement the solution. The problem is how to arm ourselves with ideas and methodologies to increase the likelihood of helping and to reduce or even eliminate the likelihood of an unsuccessful outcome.

Ed Schein has developed a 'process consultancy' model which is more complex than traditional views, attempting to fit consulting activity explicitly with real problems.

> Process consultation is a set of activities on the part of the consultant that help the client to perceive, understand, and act upon the process events that occur in the clients' environment.
>
> (Schein, volume 2, 1987, page 34.)

The focus is on how things are done, rather than on the things being done. This is in contrast to traditional models such as the purchase of expertise or the doctor-patient relationship in both of which the clients abdicate at least some of their responsibility. The manager-consultant must be willing to sit and listen, rather than to issue commands, similar to the approach known as management by walking about (MBWA).

It is assumed that the client does not necessarily understand the source of the perceived problem, though they will generally be able to describe the symptoms of the problem. The client may be aware of some of the types of help which are available, but is unlikely to know about all of them or how they might be applied to solve the problem. The aim of process consultation is therefore to help the client to learn to diagnose problems and that this is achieved, in part, through participation in the process of diagnosis. To achieve this there must be a constructive relationship between the client and the consultant.

> When individual colleagues or students came to me with problems, I found that the best stance was to keep asking questions that would clarify how the person was seeing the problem and, more important, to ask what the person had already tried to do about it. New ideas usually emerged that could be implemented. I found myself to be most effective and helpful if I took a task process orientation unless I had specific information that needed to be shared, in which case I would, of course, share it.
>
> (Schein, volume 2, 1987, page 59.)

There is a problem of the observer interfering with what he is observing; people behave differently for outsiders. Likewise, it is essential to avoid traps, to identify possible reasons for failing to see the real problems arising from:

● cultural assumptions;
● personal defensive filters and biases;
● expectations based on previous experiences.

To do this, you must identify your own biases and your own cultural assumptions and how they might affect your judgement and reason.

Implementation

The introduction of a real commitment for managing our environment requires significant expenditure in staff development and training, if organizations are to change the attitudes and ways in which people work. All too often this will be neglected, usually through 'penny-pinching', in order to achieve 'hurdle' rates of payback. Moreover, some spirit of enquiry must be present and be allowed, because managing our environment must permeate all aspects of an organization's everyday work. A considerable measure of experimentation is necessary which requires a temporary lifting of tight limits on measuring performance and a willingness to adapt. Continuous learning both formal and informal are needed, if progress is to be made up the learning curve.

The organizational reforms which are necessary if new environmental concerns and commitments are to be pursued effectively, include:

● participation;
● flexible assignment patterns;
● multiple skilling;
● self-supervision;
● quality problem-solving groups.

All of these are aimed at empowering staff, through:

- *motivation*   can job redesign and an organization's commitment to the environment be used to motivate people to achieve improved levels of performance?
- *competence*   can the levels of competence be increased through new ways of working?
- *co-ordination*   can the co-ordination between functions be improved?

The problems which must be faced include the design of:

- jobs with sufficient breadth and flexibility to allow people to understand all or large parts of the organization and their significance;
- connections between those who share responsibility, in order to improve collaboration and the management of interdependence;
- jobs that generate high-levels of motivation;
- jobs that promote learning about the system and about the business tasks themselves.

Corporate social responsibility

For managing our environment, the internal corporate culture is important. Similarly, the external corporate identity with regard to the environment is becoming of increasing importance. In the context of this discussion, many of the issues raised relate to the developing interest in corporate social responsibility, which is broader than the traditional market tests of a competitive business environment and corporate performance. The excesses of the 1980s, for instance, linked to the growing consumerism, corporate and national debt, have caused much reflection. Our environmental problems are real, complex, interrelated and long-term. Businesses (and governments) can no longer not act. New forms of relationships and responsibilities are needed. Many activities of management have a moral dimension; even if business is deemed to be about maximizing shareholders' wealth, it has an ethical underpinning. Traditional economic thinking is concerned with consumption, to maximize consumer satisfaction with an efficient allocation of resources. However, in business thinking, a broader perspective has been taken for many years. Alfred North Whitehead, for instance, in his 1929 speech at the inauguration of Harvard's Graduate School of Business Administration, argued,

> The behaviour of the community is largely dominated by the business mind.
> A great society is a society in which its men of business think greatly of their functions. Low thoughts mean low behaviour, and after a brief orgy of

exploitation, low behaviour means a descending standard of life. The general greatness of the community, qualitatively as well as quantitatively, is the first condition for steady prosperity, buoyant, self-sustained, and commanding credit.

While economic self-interest will remain an important force for managers and their organizations and different objectives will remain for business managers and different constituencies, as we approach the next millennium, it is becoming more likely that different interest groups will increasingly gain the power to influence the markets by their values. An unstable and unreliable relationship exists between economic and ethical valuation. A new, concerned style of management is likely to emerge, as managers are expected to serve both public interest and private profit. That is, management must not be solely preoccupied with shareholders; there is a need to consider other stakeholders, and to modify objectives, such as profit satisficing rather than short-term profit maximization.

At different levels, from local communities to the support of developing countries by developed countries, the moral and political aspects of 'justice' are raised. Corporate social responsibility with regard to the environment reflects custodian and long-term interests, which may be in conflict with other objectives. This responsibility, at least at present, is largely defined and accountable to the managers (through the market), rather than society in general. Simply stated, in such circumstances, responsibility is specified as managers decide. In considering *Capitalism and Freedom*, Milton Friedman (1962, pages 133–134) questioned the principle and practice of such self-determination of corporate social responsibility by managers, considering it to be a

... fundamental subversive doctrine.

More recently, in considering *The New Realities*, Peter Drucker (1989) also queried whether 'good intentions' are sufficient. More fundamentally, it is appropriate to examine whether the so-called free competition via a market mechanism can achieve the public good, without interference from government. Conventionally, economists argue that free competition is in the public interest because the market mechanism provides:

● efficient resource allocation;
● more personal freedom.

Moreover, market failures exist because of:

● a lack of information;
● externalities (with their non-market costs).

However, this perspective focuses on the level of outputs, rather than the distribution of outputs.

Individual freedom is the main value of the market, although effective choice can only exist if there is a range of substitutable products and services available which can be afforded by the individual. While Friedrich Hayek (1976) argued that such distributional issues are separate considerations, interestingly, Milton Friedman (1962) argued for a minimum income threshold with necessary income redistribution occuring through a negative income tax. Philosophically, the underlying proposal is for some justice in the market mechanism, an egalitarian minimum, although not necessarily equality – 'fairness'. For example, Ernest van den Haag (1976, page 109) suggested,

> Justice is as irrelevant to the functioning of the market, to economic efficiency, and to the science of economics, as it is to computers or to the science of meteorology. But it is not irrelevant to our attitude toward these things. People will tolerate a social or economic system, however efficient, only if they perceive it as just.

In a recent consideration of business systems and their values, James Kuhn and Donald Shriver (1991, pages 201–202 and 205–206) summarize social responsibilities for today's managers as *Beyond Success* by stating,

> Not only do managers prefer fair to free competition, profit satisficing to profit maximizing, and effective to efficient performance, but they have also discarded or reinterpreted many of the other characteristics and values that economists find in the market system. They apply the adjective 'free' to their enterprises and an absence of government regulation rather than to the market. They are ambivalent about whether a free market unregulated by government for 'proper' probusiness policies can help them greatly. They do not gain say subsidies, protective legislation, or tax advantages. Fair competition is better than free competition because a more planned, controlled and manageable market is preferable to the harsh uncertainties and instabilities of the market. Likewise managerial effectiveness – the ability to meet self-defined goals and to serve self-proclaimed purposes – is more certain of achievement than is the pursuit of market efficiency.

That is,

> The business system, as managers practise and interpret it, is quite different from the market system that economists tout. ... In meeting consumers through the market, managers argue for fair, not free, competition. In dealing with shareholders, they seek satisfactory, not efficient (or maximum), returns. In evaluating themselves and their subordinates, they aim for effective and (but sometimes rather than) efficient performance. .... The

market envisions only such managerial right as may be won from the voluntary agreement of individuals, free to accept or reject terms at any time.

In a discussion of greener marketing, as portrayed in Table 7.1, Martin Charter (1992, page 142) summarizes the various stakeholder groups and the particular needs.

Table 7.1   Key Stakeholder Needs

Stakeholder groups	*Needs*
Customers	High quality, greener products, reasonable prices
Investors	Profitability, growth, good management
Parent company	Profitability, growth, good management
Directors	Job security, corporate image, job satisfaction
Employees	Job security, corporate image, job satisfaction
Community	Social responsibility, security, open dialogue
Legislators	Social responsibility, open dialogue, over-compliance
Pressure groups	Social responsibility, open dialogue, environmental excellence
Suppliers	Secure contracts, long-term relationships, growth
Media	Good news, bad news, open dialogue

Case: The Body Shop: Body and soul

The first Body Shop opened in 1976 in Brighton. Today, The Body Shop is a global business (on a franchise basis) with more than 800 shops in over 40 countries. The company is probably best-known for its personal values and policies on reusable and recyclable packaging, employee training and education, and its policy against animal testing.

For The Body Shop, marketing is not about selling shampoos and body lotion, but about creating a reliable corporate image founded on a belief in human values and social responsibility. Protection of the environment, involvement with local communities and lobbying for changes in social values are the wider issues that are deeply rooted in the company's culture. There is no doubt that The Body Shop's founder, Anita Roddick, is motivated by altruism, and the way in which The Body Shop has managed to successfully communicate their message and build credibility throughout the world is untraditional. The company has an Environmental Projects Department (since 1986), which monitors all practices and products of the company and makes sure that they accord with the underlying principles of the organization and its founder.

Campaigning is focused around social and environmental matters. The Body Shop has made joint poster campaigns with for instance Friends of the Earth and Greenpeace on acid rain, recycling, ozone layer, and green consumers. But today the shop windows are the only form of advertising, accompanied by posters, leaflets, messages on carrier bags and staff involvement. The company promotes two major international campaigns each year: one on human rights and one on the environment.

Roddick's explanation of the connection between caring for the environment and selling moisture cream is a useful illustration of the organization's philosophy and practices,

> First, our environmental campaigning raised the profile of the company considerably, attracted a great deal of media attention and brought more potential customers into our shops. On that basis alone it could be justified as a sensible commercial decision. And much more important, in my view, was the tremendous spin-off for our staff, enabling them to get involved in things that really matter – pushing for social change, improving the lot of the underprivileged, helping to save the world.

In 1991, The Body Shop won one of 'The Better Environment Awards' (which was sponsored by the Environment Foundation, Department of the Environment and by Shell UK) for incorporating environmental concerns into every aspect of its operations. The same year the company carried out an internal environmental audit of its operations and planned to commission an external 'green' audit. The organization structure of the company was reorganized and each department has now an environmental adviser who works with the Environmental Department.

The Body Shop serves as a prime example of how sound (and unusual) environmental selling can be incorporated into the culture of an organization, resulting in both good corporate citizenship and a healthy and profitable business. Because the above-mentioned principles have guided the company's way of making business from the very outset, they have become a natural ingredient of The Body Shop concept. Success and social responsibility, even in relation to selling, are not incompatible.

(Source: Roddick, 1991.)

Postscript:
The environmental commitment has been reemphasized, as investors indicate worries over The Body Shop's share price slump.

Concluding comments

For managing our environment, the issue for managers is not whether they can cope with change, but how; for there is no alternative. The first thing to do is to set an example. An important aspect of leading is the ability to empower, not simply to delegate the dirty or the dull jobs but to take the risk of a subordinate proving that he or she is better than you are!

Peter James (1992) suggests that effective corporate responses to the environment require organizations to become 'CLEANER', and he explores the implementation through:

- Coherence;
- Leadership;
- Enthusiasm;
- Awareness;
- New processes and products;
- Efficiency;
- Review.

Managing change cannot be separated from the organizational culture, which contains within it many of the determinants of success and failure. Consequently, we must look carefully at the processes being used to consult, to manage and to implement. They must take into account the culture of the organization, both because the processes should reflect how things are done in the organization and because they should also warn us when the culture must be altered.

Recognizing that the success of an organization, especially in the longer term, depends on the abilities of people, whether it is termed intellectual capital or core competencies, it is essential that recruitment is meticulous and that people are subsequently trained extensively. It is important to build links to other organizations whose intellectual capital is of value, for example, consultancies, advertising agencies and research laboratories. The overall aim is to create a 'learning organization'; an organization which can learn faster and better than its rivals and which can turn that knowledge into products or services or improvements in product or service characteristics which will win competitively. Such an organization creates transformation and adapts to the various, external pressures, including the growing concern for our environment.

Review questions

1. Consider the assertion that successful environmental management requires commitment by all people in an organization.
2. How do you expect environmental management to change the ways in which managers will operate in the next five to ten years? What skills and knowledge will you need to acquire as a consequence of these changes?
3. What are the advantages of process consultancy over more traditional forms of consultancy for managing our environment?

Study questions

1. From observations in an organization, examine how decisions affecting the environment are taken by individual managers.
2. How would you make your organization more environmentally oriented?
3. Discuss the statement that its people are an organization's most important asset.

Further reading

Beer, Michael, Spector, Bert, Lawrence, Paul, Mills, Quinn and Walton, Richard (1984) *Managing Human Assets*, Free Press, New York.

A classic text from the Harvard Business School on human resources management. Although published in 1984, it remains as valuable today as when published, providing views which link HRM to the strategic management of organizations.Belasco, James A. (1990) *Teaching the Elephant to Dance*, Hutchinson Business Books, London.

This book is very useful for the practitioner. It is unlikely to be read from cover to cover, but can easily be dipped into either for specific issues, provocative questions or to find new and interesting ideas and, most importantly, suggestions about how to employ those ideas.Crozier, Michel (1990) *L'Organisation à l'Ecoute*, Les Éditions d'Organization, Paris.

In French. A leading French industrial sociologist's view of HRM. It demonstrates some of the differences between French management tradition and the Anglo-American approach.

Davis, John (1991) *Greening Business*, Basil Blackwell, Oxford.

Within an important series on Developmental Management and influenced primarily by the ideology and argument of Fritz Schumacher, John Davis takes a holistic view, asking for radical changes in our present economic system in order to create a

... new path of discriminating 'sustainable development'.

This, it is argued, has become an urgent imperative for business as the industrial development within the last two centuries has resulted in systematic damage to the planet – a situation which will have to be stopped now, before it is too late. Davis points out the responsibility of the private business sector in this 'revolution' and calls attention to the importance of changes in our basic beliefs and values and consequently in the organization of businesses and our society as a whole. The emphasis is on a shift in focus from a money-driven value system to a people-centred approach.

Moss Kanter, Rosabeth, Stein, Barry A. and Jick, Todd D. (1991)
The Challenge of Organizational Change, Free Press, New York.

This book addresses a fundamental management issue for the 1990's of how to face change. Case studies are used to support the arguments and, interestingly, they show both how to introduce change and how not to introduce change.

Quinn Mills, D. (1991) *Rebirth of the Corporation*, John Wiley and Sons, New York.

With obvious linkages to total quality management, striving for continuous improvement, Quinn Mills argues for organizational transformation, replacing the hierarchical organization by the cluster organization. The orientation of the clusters' development are relationships with customers and suppliers and information requirements – that is strategy, rather than structure. An interesting example of successful clusters is provided by the information services of Du Pont Fibers.

Simply stated, an excellent, thought-provoking volume.

Schein, Ed (1987) *Process Consulting*, (two volumes), Addison-Wesley, Reading.

A very useful book for the consultant and practitioner alike. It identifies ways in which it is possible to work with groups of people and help them solve their own problems.

Schein, Ed (1989) *Organisational Culture and Leadership*, Jossey-Bass, San Fransisco.

Rather heavier going than *Process Consulting*, this is the academic side of organizational culture, primarily for those interested in pursuing the theoretical basis of the ideas.

8 Towards the next millennium

In our day, there is a growing awareness that world peace is threatened not only by the arms race, regional conflicts and continued injustices among people and nations, but also by a lack of due respect for nature, by the plundering of natural resources and by a progressive decline in the quality of life.

His Holiness Pope John Paul II
World Peace Day
1 January, 1990

Introduction

Arguing for new approaches to the environment and development, the Norwegian Prime Minister Gro Harlem Brundtland (1987, page 37) stresses that,

> Failures to manage the environment and to sustain development threaten to overwhelm all countries. Environment and development are not separate challenges; they are inexorably linked. Development cannot subsist upon a deteriorating environmental resource base; the environment cannot be protected when growth leaves out of account the costs of environmental destruction. These problems cannot be treated separately by fragmented institutions and policies. They are linked in a complex system of cause and effect.

Moreover, in his Inaugural speech on 20 January, 1993, US President Bill Clinton argues that,

> To renew America, we must meet challenges abroad as well as at home. There is no longer clear division between what is foreign and what is domestic – the world economy, the world environment, the world AIDS crisis, the world arms race – they affect us all. As an old order passes, the world is more free but less stable.

His Vice-President has a more explicit focus on the environment.

> For civilisation as a whole, the faith that is so essential to restore the balance now missing in our relationship to the earth is the faith that we do have a future. We can believe in that future and work to achieve it and preserve it, or we can whirl blindly on, behaving as if one day there will be no children to inherit our legacy. The choice is ours; the earth is in the balance.
>
> Vice-President Al Gore (1992, page 368.)

An increasing number of businesses recognize their important role. For instance,

> We are . . . striving to make BP an industry leader in HSE (health, safety and environmental) performance. This makes sense not just from a moral standpoint but also as a matter of sound business practice. Today, a good HSE performance is an integral part of efficient and profitable business management. However, for a global company like BP, the aim of being an industry leader in HSE is far from straightforward, because of the differing standards and expectations from country to country. Yet this can also provide an opportunity to secure a competitive edge – for example, by transferring technologies from regions with strict environmental standards to those whose standards are currently less stringent.
>
> (1991 BP Annual Report.)

Stephan Schmidheiny (1992, page xxii) describes why the global business perspective on development and the environment by the Business Council for Sustainable Development is entitled *Changing Course*.

> While the basic goal of business must remain economic growth, as long as world population continues to grow rapidly and mass poverty remains widespread, we are recommending a different course toward that goal. There will be changes in direction and changes in the measurements of progress to include indicators of quality as well as quantity. Business is a large vessel; it will require great common effort and planning to overcome the inertia of the present destructive course, and to create a new momentum toward sustainable development.

The progress towards sustainable development has direct implications for the future of our planet and human civilization. At a global scale, many environmental problems cannot be separated from the unequal power and prosperity of developed and developing countries. In looking forward to the next millennium, our vision and planning must take on new horizons, explicitly recognizing our citizenship in a global village. As we move further into our knowledge-based society and economy, people become more and more the main real asset or resource available to ensure continued progress. Knowledge, especially access to knowledge, will increasingly have an important role with regard to power, which is inherent in all social organizations and systems. As Alvin Toffler (1991, page 11) argues in his vision from the edge of the twenty-first century,

For we stand at the edge of the deepest powershift in human history.

It is correct to argue that all human activity affects the environment. However, this does not mean all activities should be detrimental. Indeed, with appropriate policies and management, humans should be able to provide added value by creating more stable, productive and sustainable environments. In looking to the future for managing our environment, in this final chapter, we broaden the discussion in a social, political and global context. Specifically, attention is given to three important issues:

● demographics, particularly population growth and distribution;
● poverty, particularly the relationships with the environment;
● technology, particularly the links with international trade and technology transfer.

Although the current scale of the population and its forecast expansion in developing countries is a major issue, with regard to environmental concerns, this problem cannot be separated from their problems of poverty and their lack of appropriate technology platforms. This situation has led to a short-term exploitation of local, natural resources. In addition, a brief description of US Vice-President Al Gore's so-called 'Global Marshall Plan' is presented, and, in the final section, the business case for managing our environment is summarized.

Case: Global environmental complexities: Avoid common pitfalls

Quite understandably, the environment is a matter that is considered passionately and from a number of different ideological stances. It is, however, an extremely complicated topic that cannot be divorced from its interrelationships with socioeconomic and political issues at different, local to global scales. Pitfalls can arise without clear thinking. Donella Meadows and her colleagues (1992, pages 229–230) highlight some common misconceptions, including:

● **Not**: A warning about the future is a prediction of doom;
● **But**: A warning about the future is a recommendation to follow a different path.

● **Not**: The environment is a luxury or a competing demand or a commodity that people will buy when they can afford it;
● **But**: The environment is the source of all life and every economy.

● **Not**: Change is sacrifice;
● **But**: Change is challenge and it is necessary.

● **Not**: Stopping growth will lock the poor in their poverty;

- **But**: Present patterns of growth are locking the poor into poverty; they need growth that is specifically geared to serve their needs.

- **Not**: Everyone should be brought up to the material level of the richest countries;
- **But**: All material human needs should be met materially and all nonmaterial needs met nonmaterially.

- **Not**: All growth is good, without question, discrimination, or investigation;
- **And not**: All growth is bad;
- **But**: What is needed is not growth, but development. Insofar as development requires physical expansion, it should be equitable, affordable and sustainable.

- **Not**: Technology will solve all problems, or technology does nothing but cause problems;
- **But**: What technologies will reduce throughput, increase efficiency, enhance resources, improve signals, end poverty, and how can society encourage them?
- **And**: What can we bring to our problems as human beings, beyond our ability to produce technology?

- **Not**: The market system will automatically bring us the future we want;
- **But**: How do we use the market system, along with many other organizational devices, to bring us the future we want?

- **Not**: Industry is the cause of all problems, or the cure;
- **Nor**: Government is the cause or cure;
- **Nor**: Environmentalists are the cause or the cure;
- **Nor**: Any other group (economists come to mind) is the cause or cure;
- **But**: All people and institutions play their role within the large system structure. In a system that is structured for overshoot, all players will deliberately or inadvertently contribute to that over-shoot. In a system that is structured for sustainability, industries, governments, environmentalists, and most especially economists, will play essential roles in contributing to sustainability.

- **Not**: Unrelieved pessimism;
- **and not**: Sappy optimism;
- **But**: The resolve to discover and tell the truth about the successes and the failures of the present and the potentials and the obstacles in the future;
- **And above all**: The courage to admit and bear the pain of the present world, while keeping a steady eye on a vision of a better future.

If the problems of the developing countries are not addressed by everyone, it is now clear that a sustainable future for the developed world will be unrealizable. Although action is likely to require local initiatives, international co-operation is an essential prerequisite (see Table 8.1). The Business Council for Sustainable Development concludes,

> The concept of sustainable development offers guidelines but no model. Its main truth is that economic progress, social progress, and the sound management of environmental resources must all proceed apace. To strive for the first goal while ignoring the second two destroys the basis of all progress. ... The true global business challenge ... is to benefit from the system while contributing to and improving it. This is the essence of sustainable development.
>
> (Schmidheiny, 1992, pages 161 and 179.)

Table 8.1 Summary of the Brundtland (1987, pages 348–351) Report's Proposed Legal Principles for Environmental Protection and Sustainable Development

---

I GENERAL PRINCIPLES, RIGHTS, AND RESPONSIBILITIES

Fundamental human right
  1. All human beings have the fundamental right to an environment adequate for their health and well-being.

Inter-generational equity
  2. States shall conserve and use the environment and natural resources for the benefit of present and future generations.

Conservation and sustainable use
  3. States shall maintain ecosystems and ecological processes essential for the functioning of the biosphere, shall preserve biological diversity, and shall observe the principle of optimum sustainable yield in the use of living natural resources and ecosystems.

Environmental standards and monitoring
  4. States shall establish adequate environmental protection standards and monitor changes in and publish relevant data on environmental quality and resource use.

Prior environmental assessment
  5. States shall make or require prior environmental assessments of proposed activities which may significantly affect the environment or use of a natural resource.

Prior notification, access and due process
  6. States shall inform in a timely manner all persons likely to be significantly affected by a planned activity and to grant them equal access and due process in administrative and judicial proceedings.

Table 8.1   (*continued*)

Sustainable development and assistance
   7. States shall ensure that conservation is treated as an integral part of the planning and implementation of development activities and provide assistance to other States, especially to 'developing' countries, in support of environmental protection and sustainable development.

General obligation to co-operate
   8. States shall co-operate in good faith with other States in implementing the preceding rights and obligations.

II   PRINCIPLES,   RIGHTS,   AND   OBLIGATIONS   CONCERNING TRANSBOUNDARY   NATURAL   RESOURCES   AND   ENVIRONMENTAL INTERFERENCES

Reasonable and equitable use
   9. States shall use transboundary natural resources in a reasonable and equitable manner.

Prevention and abatement
   10. States shall prevent or abate any transboundary environmental interference which could cause or causes significant harm (but subject to certain exceptions provided for in 11 and 12 below).

Strict liability
   11. States shall take all reasonable precautionary measures to limit the risk when carrying out or permitting certain dangerous but beneficial activities and shall ensure that compensation is provided should substantial transboundary harm occur when the activities were not known to be harmful at the time they were undertaken.

Prior agreements when prevention costs greatly exceed harm
   12. States shall enter into negotiations with the affected State on the equitable conditions under which the activity could be carried out when planning to carry out or permit activities causing transboundary harm which is substantial but far less than the cost of prevention. (If no agreement can be reached, see 22).

Non-discrimination
   13. States shall apply as a minimum at least the same standards for environmental conduct and impacts regarding transboundary natural resources and environmental interferences as are applied domestically (that is, do not do to others what you would not do to your own citizens).

General obligation to co-operate on transboundary environmental problems
   14. States shall co-operate in good faith with other States to achieve optimal use of transboundary natural resources and effective prevention or abatement of transboundary environmental interferences.

Table 8.1   (*continued*)

Exchange of information
15. States of origin shall provide timely and relevant information to the other concerned States regarding transboundary natural resources or environmental interferences.

Prior assessment and notification
16. States shall provide prior and timely notification and relevant information to the other concerned States and shall make or require an environmental assessment of planned activities which may have significant transboundary effects.

Prior consultations
17. States of origin shall consult at an early stage and in good faith with other concerned States regarding existing or potential transboundary interferences with their use of a natural resource or the environment.

Co-operative arrangements for environmental assessment and protection
18. States shall co-operate with the concerned States in monitoring, scientific research and standard setting regarding transboundary natural resources and environmental interferences.

Emergency situations
19. States shall develop contingency plans regarding emergency situations likely to cause transboundary environmental interferences and shall promptly warn, provide relevant information to and co-operate with concerned States when emergencies occur.

Equal access and treatment
20. States shall grant equal access, due process and equal treatment in administrative and judicial proceedings to all persons who are or may be affected by transboundary interferences with their use of a natural resource or the environment.

_____

III STATE RESPONSIBILITY
21. States shall cease activities which breach an international obligation regarding the environment and provide compensation for the harm caused.

_____

IV PEACEFUL SETTLEMENT OF DISPUTES
22. States shall settle environmental disputes by peaceful means. If mutual agreement on a solution or on other dispute settlement arrangements is not reached within 18 months, the dispute shall be submitted to conciliation and, if unresolved, thereafter to arbitration or judicial settlement at the request of any of the concerned States.

Demographics

The population issue has significant economic, ethical, racial, social and political dimensions. Rapid population growth obviously places enormous additional demands on resources – education, food, health care, housing, and so on, even without the desired increased consumption levels for many people living in developed countries. It can be argued that this is the major challenge facing the world, its future development and its environment. At present, there are approximately 5 billion people on the planet. The United Nations now projects the world population will be more than 6 billion by the millennium and greater than 8 billion by 2025 and 10 billion by 2050. Associated with this population explosion is the rapid urbanization, with cities apparently attracting the poor, and its related social problems. Figure 8.1 portrays a recent estimate of the world's population growth until past the middle of the next century. With over 90 per cent expected in the developing countries of Africa, Asia and Latin America, there will continue to be difficulties to obtain basic necessities of education, employment, food, healthcare and homes.

> Slightly over one billion people, less than a quarter of the world's population, live in nations whose standard of living – health, education, diet, housing, and quantity of material possessions – has improved dramatically over what the vast majority of the world's population enjoyed a century ago. But some four billion people don't. They live in nations where average per capita wealth is only about a fifteenth of that of the rich nations and where their babies are some five to twenty times as likely to die by the age of one. Of those, nearly a billion live in 'absolute poverty' – defined as being too poor to buy enough food to maintain health or perform a job.
>
> (Ehrlich and Ehrlich, 1991, page 41)

People (billion)

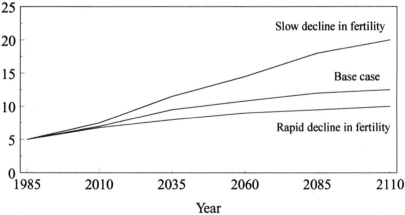

Figure 8.1 World Population Estimate

Paul Ehrlich's (1968, page 61) much-publicized *The Population Bomb* highlighted, at a global scale, the expected disaster from a continued population explosion unless it is brought under control. He argued that we are playing 'environmental roulette'. In the eighteenth century, in his *Essay on the Principle of Population*, Thomas Malthus argued that population growth would continue until it would be checked naturally through insufficient food, resulting in famine and war.

Writing over twenty years after *The Population Bomb*, Paul and Anne Ehrlich (1991, page 214), who can be described as neo-Malthusians, state that,

> Remarkably little has been accomplished in population control . . . since *The Population Bomb* appeared. Global population growth has slowed a little, but nearly all that slowdown is due to fertility reductions in two principal regions: China and the industrial nations, especially the West. A few other developing nations have achieved significant fertility declines, but most are growing as rapidly as before.

Indeed, Latin America is described as 'some hope and some horror', and Africa is thought to be 'a demographic basket case'. The Ehrlichs (1991, pages 17 and 18) connect population explosion and environmental deterioration.

> Global warming, acid rain, depletion of the ozone layer, vulnerability to epidemics, and exhaustion of soils and groundwater are all . . . related to population size.

However, they accept,

> Of course, the environmental crisis isn't caused just by expanding human numbers. Burgeoning consumption among the rich and increasing dependence on ecologically unsound technologies to supply that consumption also play major parts. This allows some environmentalists to dodge the population issue by emphasizing the problem of malign technologies. And social commentators can avoid commenting on the problem of too many people by focusing on the serious maldistibution of affluence.

It must be stressed that,

> We don't live in a surprise-free world. When *The Population Bomb* was written, we and our colleagues were enormously worried about the course that humanity was on. Yet it is sobering to recall that the book appeared before depletion of the ozone layer had been discovered, before acid precipitation had been recognized as a major problem, before the current rate of tropical forest destruction had been achieved, let alone recognized,

before the true dimensions of the extinction crisis had been perceived, before most of the scientific community had recognized the possibility of a nuclear winter, and before the AIDS epidemic.

(Ehrlich and Erhlich, 1991, page 177.)

In summary,

The evidence indicates that population growth has two broadly counter-acting impacts on environment and development. By 'forcing' adaptive and technological change, population growth may actually increase the pro-spects for development in the traditional sense of rising GNP per capita. By contributing to the depletion of natural resources - primarily the renewable open access and common property natural resources – population growth impedes development in the traditional sense and certainly reduces environmental quality. The balance of these two broad impacts favours the view that population growth, at least on the scale now being witnessed, is detrimental to future human welfare.

(Pearce, 1991, page 133).

The issue is not one of population density, but one of the 'carrying capacity' of the local environment. Simply stated, if the long-term carrying capacity of an area is being reduced by its existing population level, the area is overpopulated. Following Ehrlich and Holdren's (1974) argument, the specification of the 'impact of population growth' is presented as a function of three factors:

- P – population;
- A – affluence (a variable of resource consumption); and
- T – technology.

That is, the environmental impact, I, can be defined as:

$I = f (P,A,T)$.

Without action towards the continued population explosion, the scale of the future pressures on the planet will mean that the potential benefits of other actions may never really accrue. Although it may be difficult, if not impossible, to forecast the scale and timing of future catastrophes, it is becoming more evident that the growing pressures on the existing social and ecological systems will lead to new, much more serious, social and political difficulties.

It is noted that overpopulation exists all over the world, but the accommodation of the problems is generally more serious in developing countries. While population growth is only one of many factors behind environmental degradation, given the continuing population explosion in developing countries, continued rapid economic development is needed which will put further pressure on the planet. In fact, the Brundtland (1987, page 11) report emphasizes that,

> The issue is not just numbers of people, but how those numbers relate to available resources. .... Human resource development is a crucial requirement not only to build up technical knowledge and capabilities, but also to create new values to help individuals and nations cope with rapidly changing social, environmental, and development realities.

However, it should not be forgotten that, if developing countries are able to generate their desired economic growth, there will be an enormous increase in the consumption of resources on both an aggregate and per capita basis. In addition, it would be misleadingly simplistic if the reader is left with the impression that the population - environment interrelationship is only one of exponential growth and the (local) environment's carrying capacity. Other important and associated issues include the developing countries' modified population structure, the changing role of women and child labour, the relationships between literacy and fertility, the potential disasters from AIDS and so on.

Poverty

The problems of poverty cannot be divorced from the pressures caused by the population growth. Poverty is increasing, not decreasing. Moreover, there are important distributional and equity issues, not only between say developed and developing countries, but also within all countries. Transformations in attitudes and knowledge are essential, because of a failure really to act about the contemporary world situation in which approximately one-fifth of the world's people consume over three-quarters of its resources, while a similar proportion of the world's people are concerned exclusively with survival.

> The gross imbalances that have been created by concentration of economic growth in the industrial countries and population growth in developing countries is at the centre of the current dilemma. Redressing these imbalances will be the key to the future security of our planet – in environmental and economic as well as traditional security terms. This will require fundamental changes both in our economic behaviour and our international relations.
>
> (Strong, quoted in Schmidheiny (1992, page 3).)

As the United Nations (1991, page 28) *Human Development Report* indicates,

It is ironic that significant environmental degradation is usually caused by poverty in the South – and by affluence in the North.

The United Nations (1992) *Human Development Report* demonstrates that, over the last twenty years, the gap between the 'rich' and 'poor' countries has grown twice as wide. Willy Brandt's (1980) Commission on the *North-South* divide considered primarily education, health and housing improvements for the world's poorest people. Their focus was on resources, rather than on the environment specifically, but they did highlight the requirement for ecological stability as a condition for any long-term progress in developing countries. Following the Brundtland Report and the subsequent Earth Summit's acceptance of the concept of sustainable development, in late 1992, the United Nations and UNESCO established a World Commission on *Culture and Development* under the chairmanship of the UN's former Secretary-General, Javier Perez de Cuellar; the Commission will report in 1995. Significantly, the starting point is a belief that, development, although it requires growth, cannot be confined only to economic development, and that the key to preserving the environment is culture.

As highlighted by the recent Rio de Janeiro Earth Summit, the links between development and the environment is complex. Significantly, the problems facing the developed and developing nations are different. Moreover, the solutions are different. Generally, developed countries desires are for more consumption and developing countries' desires are for mere survival. It is, therefore, not surprising that different policies are needed. In developing countries, which are not benefiting from development and technological advancement, the policies to overcome environmental problems may need to be more related to socioeconomic development policies than to environmental issues specifically.

While the requirements of the developed and developing countries for managing our environment are different, the developed countries must remember the problems are global and interrelated; without the co-operation of the developing countries, no real progress can be made and the environmental problems cannot be resolved. It should be appreciated, for example, that developing countries are likely to value pollution control less than developed countries because of their desires and requirements for growth. Environmental policies, consequently, are extremely complicated, and the bases for different trade-offs and priorities vary. However, without rapid economic development, there is no real opportunity for the developing countries to be able to accommodate their additional people, even at today's undesirable low standards. In these countries, there is a vicious, downward spiral of population increase and poverty.

Attention here is focused briefly on some of these issues in developing countries. The Brundtland Report (1987, page 29) argues that,

> Poverty is a major cause and effect of global environmental problems. It is therefore futile to attempt to deal with environmental problems with a broader perspective that encompasses the factors underlying world poverty and international inequality.

In considering the global commons, David Pearce (1991, page 11) emphasizes that action by developed countries will be essential to obtain commitment and action by developing countries. In fact, he suggests,

> ... the rich countries of the world must pay twice: once to 'clean up' the environmental degradation they have created through environmentally insensitive growth, and twice to prevent the developing world engaging in yet further environmental destruction. Additionally, existing degradation in the developing world has to be addressed.

It is perhaps inevitable and correct that the developed countries take a lead responsibility for the global environmental challenges, because, although they do not have the majority of the world's population, they do use a disproportionate share of its resources. For instance, with the carbon dioxide emissions from fossil fuel consumption being primarily a product from the developed parts of the world, the costs of reducing the real threat of global warming should be borne by those countries that are also benefiting from the activities partly causing the problems. However, more generally, for developed countries, it may be cheaper and easier to reduce the pollution in developing countries than at home and provide benefits for everyone. As The World Bank (1992) emphasizes in its report on *Development and the Environment*, the necessary financing should not be through traditional aid but as payment for services rendered.

The symbiotic relationship between the environment and poverty are complex and the causalities are poorly comprehended and often inappropriately simplified. The two-way linkages must be considered in detail, rather than as generalities. Different people in different locations enhance or degrade their environments to varying degrees in various ways; similarly, different local environments in different locations place varying constraints and stresses on different people. The environment, like income, therefore, imposes significant influences on people's vulnerability to poverty. Moreover, it is not only the availability of resources, but also people's ability to use them effectively, that ultimately influences the extent of poverty. Consequently, perhaps not surprisingly, in developing countries, the poor, usually small farmers, are both agents and victims of environmental degradation. Environmental degradation, indeed, may impose particularly severe choices on people. Policies that are operational at a local level should consider these limitations, in the context of 'low' resource potential and 'high' resource potential, if there is to be any progress. Local, natural carrying capacity is important, although it must be appreciated that it can be changed, indeed improved, by specific actions.

The Brundtland Commission highlights the need for greater intra- and inter- generational equity, both outcomes that may be unlikely with our current economic and political system.

> The current debate over sustainable development is based on the wide-spread recognition that many investments by major financial institutions, such as the World Bank, have stimulated economic development in the Third World by encouraging the short-term exploitation of natural resources, thus emphasizing short-term cash flow at the expense of longer-term, sustainable growth. This pattern has prevailed both because of a tendency to discount the future value of natural resources and because of a failure to properly depreciate their value as they are used up in the present.
>
> (Gore, 1992, page 191.)

Economic development is a life-or-death requirement for many developing countries, and, as the promotion of peace must be alongside the promotion of the environment, specific requirements should not be imposed on these countries. However, as many of the developing countries' problems can be related to their primitive state of development, which often means few options exist other than a destruction of their natural environment, the transfer of technology is an important strategic and tactical issue.

Case: The population explosion: What can we do?

The seriousness of the current demographic trends cannot be overstressed because of their detrimental effects not only on our environment but also on our societies' structures and stability. In the final pages of their book, Paul and Anne Ehrlich (1991, pages 237–239) provide some thought-provoking 'take-home' messages,

> The 1990 population of Earth is over 5.3 billion people, and some 95 million are being added yearly.
>
> Unprecedented overpopulation and continuing population growth are making substantial contributions to the destruction of Earth's life-support systems.

Overpopulation is a major factor in problems as diverse as African famines, global warming, acid rain, the threat of nuclear war, the garbage crisis, and the danger of epidemics.

Overpopulation in rich countries is, from the standpoint of Earth's habitability, more serious than rapid population growth in poor countries.

Rapid population growth in poor countries is an important reason they stay poor, and overpopulation in those nations will greatly increase their destructive impact on the environment as they struggle to develop.

There is no question that the population explosion will end soon. What remains in doubt is whether the end will come humanely because birth rates have been lowered, or tragically through rises in death rates.

Anyone who opposes controlling the number of births is unknowingly promoting more premature deaths.

Earth cannot long sustain even 5.3 billion people with foreseeable technologies and patterns of human behaviour. if civilisation is to survive, population shrinkage below today's size eventually will be necessary.

Population control is the most pressing of human problems because of the enormous lag time between beginning an effective programme and starting population shrinkage.

A maximum number of people will live eventually if the population is reduced to a sustainable size and maintained for millions of years. Trying to see how many can live all at once is a recipe for a population crash that will lower Earth's carrying capacity and reduce the number of people that can ever exist.

The population/resource/environment predicament was created by human actions, and it can be solved by human actions. All that is required is the political and societal will. The good news is that, when the time is ripe, society can change its attitudes and behaviour rapidly.

We must all tithe to society in order to ripen the time.

Technology and trade

Broadening the discussion of the relationships between developed and developing countries, it is appropriate to comment briefly on the general issue of international aid and trade, particularly with regard to environmental issues (see also the discussions in chapter one regarding the Rio de Janeiro Earth Summit). It must be stressed that statistics highlight that the flows of (financial and/or technological) aid are not one-way to developing countries from developed countries; indeed, net inflows go to some developed countries! Moreover, it is argued by some people that much of the western aid to developing countries is a self-interested plundering which has caused environmental destruction. For example, much official aid has been directed towards large-scale and high-profile, capital intensive projects, such as dams, motorways, underground systems, and so on, many of which have had significant adverse impacts on the local environment and their communities. In addition, for example, in relation to deforestation, some developing countries, led by Malaysia, have argued that developed countries' concerns for the environment are, in part, a conspiracy to stop their economic growth – a form of 'eco-imperialism'.

Case: Trade not aid: The Body Shop approach

The Body Shop's Chairman, Gordon Roddick, argues that,

> It is these kinds of trading agreements that make our business so richly rewarding in human terms.

In proposing a new business ethic in pamphlets that are distributed to customers at their branches, The Body Shop suggests 'Trade not aid' provides a positive solution to economic hardship in the developing world. Rather than giving hand-outs, they prefer to assist local communities acquire the tools and resources they require to support themselves. It is thought that self-determination is a fundamental human right, and that this allows the developing world to retain the power to determine their own futures.

Examples of such relationships include:

- Mexico

  > In the Mesquital Valley, . . ., 600 women from three co-operatives trade with us in cactus fibre body scrubs. The women are Nanhu Indians, the fifth largest ethnic group in Mexico. Descendants of the original Aztecs, the Nanhu have cultivated maguey cactus for use as food, water, and clothing for millennia. Before making the body scrubs, the Nanhu produced craftwork for the tourist trade in Mexico City, and harvested lechugilla cactus for brush manufacturers. Both markets have collapsed. Due to lack of work in the community, many Nanhu men have left to find employment in Mexico City and the USA.
  >
  > All Trade Not Aid projects are based on the idea of harvesting only sustainable resources. The Nanhu have created a plan for the reafforestation of the maguey and The Body Shop has donated 6,000 baby cacti to help realise it. We are also looking into other products – including cloth bags, face flannels and soap mitts. Diversification of the Nanhu economy is core to their survival, and we are committed to helping them do this.

● India

> The Body Shop helped set up, and trades with, wood-turning workshops in Tamil Nadu, Southern India, which produces our popular footsie rollers and body massagers. They are made with wood from the Acacia Niloticca tree, a drought-resistant, fast growing tree which if harvested across the trunk does not need to be replanted.
>
> The project employs more than 150 craftsmen in eight workshops, providing badly needed employment in a severely depressed area. 20% of the money we pay for the rollers is committed to local welfare projects including health care, education and improved conditions at work.

In arguing for a return to basics, the pamphlet finishes with the following statement,

> We believe 'Trade not aid' is a working example of how profits and principles can go hand in hand. We are constantly in touch with primary producers and trade organisations world-wide seeking out new projects, and the next decade will see a huge growth in this area of our business. We think that trading with communities in developing countries creates a positive force they can use to shape their futures. It's back to basics in the best possible way.

In this brief section, attention is given to a number of aspects of technology, particularly,

● the appropriateness of technology;
● the economic basis for technology transfer.

As an illustration of some of the issues, special consideration is focused on agriculture and biotechnology. First, however, it is important to place this discussion into a broader context of international trade. At the time of writing, the Uruguay Round of the General Agreement on Trade and Tariffs remains unresolved. However, there are some fundamental dilemmas with regard to the GATT Agreement and the global management of our environment. For instance, Article XX assures environmental protection, but it has been in conflict with Article XI, which establishes that no trade prohibitions or restrictions can be introduced. Given the need for sustainable development, there is now an urgent requirement to amend the GATT Agreement for effective rules for environmental protection. GATT's objectives aim to discourage national standards, which can include environmental ones, when they are viewed as a disguised form of trade barrier. The interrelationships between environmental and trade policies are complex; in the European Community, the European Commission represents EC countries on trade policy, but not on environmental policy, and its two Directorates are not always in agreement!

Technologies' appropriateness is extremely important, not only because it may assist the alleviation of environmental pressures, but also because it may be the cause of new environmental pressures. For any transfer of technologies, it is essential that it is not merely a process of repeating some of the developed countries' industrialization. For instance, it is essential to avoid the transfer of products and services that cannot be both used and maintained locally. In *The Closing Circle*, for instance, Barry Commoner (1972) viewed some of the environmental problems as being the result of new technologies, his so-called 'technological flaw'. Fritz Schumacher also questioned the potential of science and technology to solve what were political and social problems, introducing the idea of 'intermediate' (or appropriate) technology.

Given the global nature of both the environmental crises and the sustainable development solutions, the policy issue of technology transfer between developed and developing countries cannot be neglected. To transfer technology from its place of origin to where it is needed is not a straightforward matter, in part because of the potential competitive advantage that can be derived from technologies' research and development (R&D) and applications which usually represent a significant and risky investment.

> The competitive advantage of an enterprise is based on technological innovation. Transferring technology means transferring competitive advantage. It is therefore a sensitive issue. Companies want expansion of business through long-term partnerships, not loss of business by the selling of their technologies.
>
> (Darwin Wika, Manager (Environment, Health and Safety), Du Pont Asia –
> quoted in Schmidheiny, 1992, page 128.)

In our increasingly technology-oriented global economy, protection of intellectual property rights is crucial. Such protection should aim to acknowledge and safeguard the creativity and development of inventions by individuals and organizations, permitting them the opportunity to benefit profitably from their time and financial resource inputs behind the research and development. Two basic issues need to be addressed:

- the harmonization, and ultimate international integration of specific, primarily national, legislation;
- the appropriate transfer of technologies between developed and developing countries.

Protection is necessary for the organizations which invest in R&D, otherwise future, much-needed developments are likely to be hindered through the lack of business incentives. On the other hand, in certain areas related to our basic humanity and to our planet, there is a cogent argument that transfer of appropriate technology should occur. The apparent dilemma of encouraging technology transfer should be resolved at a government, rather than a corporate, level. It is a problem of economic assistance, not wholly intellectual property rights, but it must also be recognized that successful technology transfer will only occur if intellectual property is protected. Software companies, for example no longer export to certain countries because their international market would become so overrun by 'legal' copies. Moreover, technical co-operation should also incorporate a priority on development of human resources, whether by acceptance of trainees and/or secondment of experts. For instance, the successes of the 'Green Revolution' were due not only to the technology, but also, and perhaps more importantly, to the integrated education programme.

The Business Council for Sustainable Development suggests that 'technology transfer' does not capture the scope and scale of the challenge raised by sustainable development. They propose the term 'technology co-operation' to capture a wider range of objectives and have a particular orientation towards business development.

> Technology co-operation concentrates on developing human resources by extending a country's ability to absorb, generate, and apply knowledge. In 'developing' countries, it works to enhance use of technology, promote innovation, and foster entrepreneurship. Technology co-operation works best through business-to-business long-term partnerships that ensure that both parties remain committed to the continued success of the project.
>
> (Schmidheiny, 1992, pages 118–119.)

The attention given to ozone depletion was the first official appreciation and action that environmental problems are not confined to national boundaries. The phased replacement of chlorofluorocarbons agreed initially in the 1987 Montreal Protocol emphasizes the need to make available to developing countries environmentally safe alternative substances and technologies. Under Article Five, signatories have agreed,

> ... facilitating access to environmentally safe alternative substances and technology

and also,

> ... facilitating bilaterally or multilaterally the provision of subsidies, aid credits, guarantees or insurance programmes

for the use of substitutes (quoted in Pearce (1991, page 69)). In addition, appreciation of the developing countries' situation led to allowance for their greater CFC use in refrigeration because of direct health benefits. The amendment agreed in London provides for a multilateral fund to assist developing countries and also highlights the need to transfer the best technologies on a fair and favourable basis.

For agriculture, some production benefits from the scientific development and use of fertilizers, pesticides and new seed varieties have accrued. The impacts of biotechnology, and its sub-discipline genetic engineering, are not concerned exclusively with agriculture. For instance, while scientific advances and their applications are difficult, if not impossible, to forecast, important impacts have been made already in waste disposal technology and recycling of some resources.

It would be incorrect to view this application of technology in some developing countries' agriculture as a complete success. New crop varieties created in laboratories, with important characteristics of consistent quality, high yields and resistance, unfortunately, often do not provide a sustainable alternative to natural strains. The long-term dangers of increasing genetic homogeneity also concern some people; indeed, some governments and private organizations are now managing gene banks to preserve natural varieties. In practice, recent increases in the world's food output have arisen because of both the expansion of the total land area that is harvested and the enhanced crop yields from modern, intensive agricultural practices. However, with the growing population, unfortunately, there is an ever increasing number of malnourished people (with an estimated 75,000 people, mainly children, dying every day). Additional, fertile land is extremely scarce, and the UN's Food and Agriculture Organization predicts future decreases in the amount of available land per person.

It is naive to assume the same technical solutions are relevant to all developing countries. Over the last quarter of a century, for example, while food production has increased in both Asia and Latin America, it has declined in Africa.

... the Asian 'Green Revolution' model is totally inappropriate for African agriculture. The Asian model had as its cornerstone improved crop varieties, which in turn were dependent on fertiliser inputs, controlled water supplies, good soil and good management. In Africa, labour is scarce, irrigation is not widespread, soils are often poor and rainfall is unreliable. Faced with an extremely sensitive environment, the African smallholder operates in a distinctive way. He is often desperately poor and therefore operates on short time horizons in a very risk-averse fashion. Any strategy aimed at mitigating rural poverty, malnutrition and environmental destruction in Africa must contain the following elements:

- a focus on smallholders, input packages and techniques (especially soil conservation techniques) based on careful on-farm research;
- community participation;
- credit facilities requiring repayment only after crop sales;
- high average returns on investment;
- agricultural price rises.

(Turner in Pearce (1991, pages 174–175.)

More generally, in the developing countries, within the Green Revolution, there has been a failure to address basic economic issues related to land ownership and distribution. It is noted that, while the focus to date in genetic engineering has been on plants, there is a developing interest with regard to animals.

International trade has always been, and will always be, an important means for countries' development. While the developed countries are now questioning the advantages of liberalized trade, it is clear that the 'developing' countries require 'free' trade for their economic development and ultimately for the protection of the environment and sustainable development. The General Agreement on Tariffs and Trade (GATT), which was established in 1948, aims by multilateral negotiations to liberalize international trade through free and fair competition in world markets, and thereby enhance economic and social growth. GATT was developed at a time prior to environmental concerns, and, therefore, it is not surprising that it is often unable directly to effectively incorporate environmental issues.

Today, a number of international trade issues are directly and indirectly relevant to the management of our environment. The Business Council for Sustainable Development highlights three significant factors:

- a rapid increase in the foreign direct investments of multinationals, which are growing much more quickly than international trade;
- a simultaneous increase and decrease in trade protectionism in the North and in the South, respectively;
- growing, ideological and substantive, conflicts between international trade policies and environmental policies.

An increasingly important business phrase is 'think globally, act locally' ('GLOCAL'). With many multinationals restructuring as local companies, advantages of more informed environmental management in developed countries can still be transferred to developing countries where there is now a relatively large degree of local autonomy and indigenous management. As countries, for example, introduce new regulations and standards regarding the environment, it can be envisaged that, in some circumstances, these new measures could represent non-tariff barriers to international trade. Instantaneous changes or rapid phasing of changes are not feasible, and probably unfair, given countries' different starting points. The extent of co-ordination by international agreement, within or outside GATT, will be an important determinant of progress regarding the complex, interrelationships between international trade and sustainable development.

Although there are obviously disadvantages of free trade, the significant benefits for countries and organizations of their comparative advantages remain and these should be exploited. However, it would be wrong to attempt to simply solve some of the environmental issues, such as internalizing costs, through mechanisms and procedures of international trade. Solutions to these problems should remain the responsibility of individual governments. However, if some countries are willing and able to internalize more environmental costs, which would be reflected in the prices of their exports, it still remains likely that, unfortunately, other countries would be willing to disregard such costs because of the opportunities for short-term profits.

Without arguing that technology will be our saviour, it is clear that appropriate technology, if managed effectively, can assist many organizations and countries. In the global context, technology transfer to developing countries is an important economic, political, and business issue. The Business Council for Sustainable Development believes,

> Long-term business-to-business partnerships and direct investment provide excellent opportunities to transfer the technology needed for sustainable development from those who have it to those who require it. This new concept of 'technology co-operation' relies principally on private initiatives, but it can be greatly enhanced by support from governments and institutions engaged in overseas development work.
>
> (Schmidheiny, 1992, page xii.)

A Global Marshall Plan

There is one Earth and, ultimately, the interlocking environmental crises are everyone's problems and they require co-ordinated, global action. Disappointingly, at present, the world vision and commitment is still lacking. Interestingly, Peter Drucker (1989, page 130) views that,

> The nineteenth century cured two of mankind's oldest scourges by transnational action – the slave trade and piracy on the high seas. It declared both to be common enemies of humanity, the suppression of which was in the interest of any country at any time. The threat to the human habitat, the ecology, is a recent threat. But it is a greater threat than the slave trade or piracy ever were, and a threat to everyone. If it can be averted at all, it can be averted only by transnational commitment and joint action.

In a political sense, the environment may now have a 'White House' effect, because of Senator, now Vice-President, Al Gore's known commitment to the improvement of the environment. At the time of writing, for example, there are clear indications that President Clinton will reverse the decision by President Bush, and sign the biodiversity treaty from the Rio Earth Summit.

Al Gore (1992, page 15) has proposed 'A Global Marshall Plan' in his consideration of the *Earth in the Balance*.

> The integrity of the environment is not just another issue to be used in political games for popularity, votes, or attention. And the time has long since come to take more political risks – and endure much more political criticism – by proposing tougher, more effective solutions and fighting hard for their enactment.

The term Global Marshall Plan is used by Gore (1992, page 297) to highlight the urgency and scale of our environmental problems.

> The scope and complexity of this plan will far exceed those of the original (Marshall Plan); what's required now is a plan that combines large-scale, long-term, carefully targeted financial aid to developing nations, massive efforts to design and then transfer to poor nations the new technologies needed for sustained economic progress, a world-wide programme to stabilize world population, and binding commitments by industrial nations to accelerate their transition to an environmentally responsible pattern of life.

Al Gore specifies five strategic goals with associated sets of policies for implementation (including some specific roles for the United States). In summary, they are:

- *the stabilizing of world population:*
  - allocate resources to fund carefully targeted literacy programmes keyed to every society where the demographic transition has yet to occur;
  - develop effective programmes to reduce infant mortality and ensure the survival and excellent health of children;
  - ensure that birth control devices and techniques are made ubiquitously available along with culturally appropriate instruction;

- *the rapid creation and development of environmentally appropriate technologies* through a world-wide Strategic Environment Initiative (SEI) (a 'crash' programme equivalent to the Strategic Defence Initiative (SDI)):
  - tax incentives for the new technologies and disincentives for the old;
  - research and development funding for new technologies and prospective bans on the old ones;

- Government purchasing programmes for early marketable versions of the new;
- the promise of large profits in a market certain to emerge as older technologies are phased out;
- the establishment of rigorous and sophisticated technology assessment procedures, paying close attention to all the costs and benefits – both monetary and ecological – of the new proposed substitute technologies;
- the establishment of a network of training centres around the world, thus creating a core of environmentally educated planners and technicians and ensuring that the developing nations will be ready to accept environmentally attractive technologies and practices;
- the imposition of export controls in developed countries that assess a technology's ecological effect;
- significant improvements in the current patchwork of laws, especially in those countries that now effectively fail to safeguard the rights of inventors and developers of new technology;
- better protection for patents and copyrights, improved licensing agreements, joint ventures, franchises, distributorships, and a variety of similar legal concepts;

● *a comprehensive and ubiquitous change in the economic 'rules of the road' by which we measure the impact of our decisions on the environment:*
   – the definition of GNP should be changed to include environmental costs and benefits;
   – the definition of productivity should be changed to reflect calculations of environmental improvement or decline;
   – Governments should agree to eliminate the use of inappropriate discount rates and adopt better ways to quantify the effects of our decisions on future generations;
   – Governments should eliminate public expenditures that subsidize and encourage environmentally destructive activities;
   – Governments should improve the amount and accuracy of information on the environmental impacts of products and provide it to consumers;
   – Governments should adopt measures to encourage full disclosure of companies' responsibility for environmental damage;
   – Governments should adopt programmes to assist companies in the study of the costs and benefits of environmental efficiency;
   – nations should revise their antitrust laws to encompass environmental harm;
   – Governments should require the incorporation of standards to protect the environment in treaties and international agreements, including trade agreements;

– environmental concerns should be integrated into the criteria used by international finance institutions for the evaluation of all proposed grants of development funds;
– Governments should make accelerated use of debt-for-nature swaps to encourage environmental stewardship in return for debt relief;
– Governments should develop an international treaty establishing limits on carbon dioxide emissions by country and a market for the trading of emission rights among countries that need more and countries that have an excess amount.

● *the negotiation and approval of a new generation of international agreements;*
● *the establishment of co-operative plan for educating the world's citizens about our global environment.*

The integrating goal is:
● *the establishment, especially in the 'developing' world, of the social and political conditions most conducive to the emergence of sustainable societies.*

A central theme of this book is the need for business action, recognizing their business opportunities and responsibilities. However, without the political will, management of our environment will remain an elusive objective.

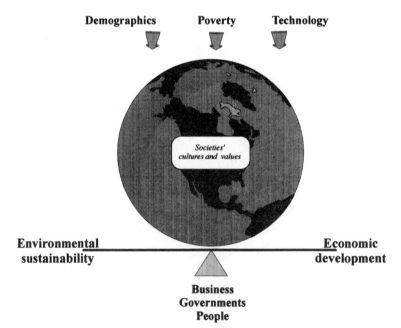

Figure 8.2   Our Single Planet

Future outlook: business opportunities and responsibilities

Since the beginning of humankind, approximately one hundred thousand years ago, the world has changed enormously, especially by our hands in recent times. There is no longer debate about the existence of interlocking environmental crises and that 'business-as-usual' is sustainable. Business is both a major cause of the problems, and their main source of solutions. Business has a responsibility to act, and, by so doing, can realize new business opportunities.

The greater incidence of an environmental consciousness among business people in recent years gives some signals for hope, but there will not be any real progress until it is translated into action. While the complex environmental pressures require co-ordinated policy and action, the decisive forces must be driven by business. The actions of businesses will determine to a large extent the achievement of sustainable development in the future. As this book has demonstrated, for managing our environment, the nature of the business opportunities and responsibilities vary in both scope and scale. As stressed from the outset of this book, managing our environment will require an interdisciplinary and multifunctional outlook, with co-operation between business, government and the general public at local, national and international levels. Isolated, disciplinary or functional perspectives are inappropriately myopic and unlikely to provide the necessary foundation for progress.

It is perhaps not an over exaggeration to state that sustainable development, like the Agricultural Revolution and the Industrial Revolution, will need a 'management revolution'. Management must become environmentally conscious through their actions, which requires:

- the establishment of environmental goals and performance;
- the introduction of data management systems to provide the necessary information to monitor performance against stated objectives;
- the allocation of resources to continually improve environmental performance;
- the specification of management responsibilities, processes and procedures to ensure the appropriate action occurs.

Potential new costs and liabilities, in a period of heightened business competition, mean that management can no longer ignore the real costs of their operations; 'more for less' may become a significant requirement. The traditional management concerns for efficiency is neither inconsistent nor in conflict with environmentally conscious management. This situation will become more true, not less, as environmental costs are recognized and internalized fully.

Comprehension and control of costs are essential for all organizations. Environmentally conscious management can provide real opportunities to reduce costs, whether they are in terms of raw materials, including energy, packaging, waste production or reduced insurance costs. Moreover, there can be new opportunities for revenue generation through new products and services in existing and new markets. For instance, in Steve Bennet's (1992) *Ecopreneuring*, he offers a guide to green opportunities in a range of industries, not only traditional recycling and waste treatment, but also in entertainment and travel. John Elkington and Peter Knight (1991, page 32) consider 'spotting the green opportunity' in *The Green Business Guide*. While highlighting the obvious growth areas of energy conservation, pollution control, waste management, water treatment, and so on, following a British Steel workshop, they highlight the different kinds of opportunities that can exist:

- to save money;
- to make money;
- to develop cleaner, quieter, faster, cheaper, better technologies and products;
- to develop and deploy new technologies;
- to get into new markets early;
- to incorporate quality;
- to win friends and influence people;
- to recruit and hold the best people;
- to develop a competitive edge; and by no means least important,
- to help save the planet.

For the environment's timescales, a significant business challenge is innovation. Following recent experience, successful developments are more likely to be market-led, than science-led, and, while investments by business will obviously be an important driving force, the significance of the national position, particularly the Government's commitment, will also be an important determinant.

To finish, it is sufficient to state that, although many of today's environmental problems have been caused directly or indirectly by business activities, it is also true that wealth creation by businesses is needed to provide the required future investment for our environment. Managing our environment represents sound business practice.

Review questions

1. Living on our single planet, consider the different pressures on developed and developing countries.

2. Consider the world's population growth, and specify some of the necessary actions for meaningful development.
3. For managing our environment, examine the role of technology, particularly the transfer between developed and developing countries.

Study questions

1. Describe the world in which you would like to live when you are seventy years old.
2. To what extent can we rely on some form of 'technological fix' to solve today's environmental crises?
3. In your organization, prioritize ten action points to enhance the management of our environment.

Further reading

Barrett, Brendan F. D. and Therivel, Riki (1992) *Environmental Policy and Impact Assessment in Japan*, Routledge, London.

Brendan Barrett and Riki Therivel provide a clear account of industrial development and environmental awareness in modern Japan, and there are many important lessons for industrial development in other countries. For example, while the pollution problems resulted in local citizen action, no real national movement has been forthcoming. Moreover, while Japan's environmental impact assessment (EIA) process was established in the 1970s following the US model, there have been real difficulties in transferring environmental legislation from one country to another with a different social, economic and political context and power relationships.

Interesting case studies of Japan's environmental policy in action cover:

● the Honshu-Shikoku bridge;
● the Kansai international airport;
● Kyoto's second outer circular highway;
● the new Ishigaki airport;
● the trans-Tokyo Bay highway.

Gore, Al (1992) *Earth in the Balance*, Earthscan, London.

A book by the most eminent Western politician that has demonstrated an interest in and commitment to improving the management of our environment. A clear, indeed eloquent, personal discussion of the threats from our environmental crises.

Al Gore argues for radical thinking and action, outlining the strategic goals and associated sets of policies for A Global Marshall Plan. Now as the Vice-President of the United States, it will be interesting to see commitment and actions of the world's most powerful nation to managing our environment.

King, Alexander and Schneider, Bertrand (1991) *The First Global Revolution*, Simon and Schuster, London.

The latest report by The Council of The Club of Rome, and a book that is subtitled, 'The world twenty years after *The Limits to Growth*' (see also chapter one).

Interdependence of problems is the main characteristic of this interesting, but uneven, account. The authors summarize the key issues as:

● the pressing need for the world to convert from a military to a civil economy, exemplified by the first steps taken to end the cold war;
● the urgent need to contain global warming by reducing carbon dioxide emissions, encouraging tropical reforestation and conserving traditional forms of energy while developing alternatives;
● the necessity of recognizing the disastrous short-term effects of exploitation by developed countries of developing countries' poverty and need.

The call for world solidarity is clear, but, even a convinced reader is left wondering what it really means and how we can attempt to achieve it.

Pearce, David W. (1991) (ed) *Blueprint 2*, Earthscan, London.

The sequel to the important and influential *Blueprint for a Green Economy*, which extends the national (British) discussion to examine the greening of the world economy. Again, another clear and comprehensive discussion of the issues, covering:

● the global commons;
● global warming: the economics of a carbon tax;
● global warming: the economics of tradable permits;
● economics and the ozone layer;
● environmental degradation in the third world;
● population growth;
● tropical deforestation;
● environmentally sensitive aid;
● conserving biological diversity;
● environment, economics, and ethics.

Bibliography

Abrahamson, D. E. (ed) (1989) *The Challenge of Global Warming*, Island Press, Washington DC.

Adams, R., Carruthers, J. and Hamil, S. (1991), *Changing Corporate Values*, Kogan Page, London.

Adams, R., Carruthers, J. and Fisher, C. (1991) *Shopping for a Better World*, Kogan Page, London.

Adams W. M. (1990) *Green Development*, Routledge, London.

Advisory Committee on Business and the Environment (1992) *Second Progress Report to and Response from The President of Board of Trade and the Secretary of State for the Environment*, Department of the Environment and the Department of Trade and Industry, London.

Ahmad, Y. J., El Serafy, S. and Lutz, E. (eds) (1989) *Environmental Accounting for Sustainable Development*, World Bank, Washington DC.

Alfsen, K. H., Bye, T. and Lorentsen, L. (1987) *National Resource Accounting and Analysis*, Central Bureau of Statistics of Norway, Oslo.

Anderson, K. and Blackhurst, R. (eds) (1992) *The Greening of World Trade Issues*, Harvester-Wheatsheaf, Hemel Hempstead.

Anderson, V. (1991) *Alternative Economic Indicators*, Routledge, London.

Andrews, K. R. (ed) (1989) *Ethics in Practice*, Harvard Business School Press, Cambridge.

Angell, D. J. R., Cromer, J. D. and Wilkinson, M. L. N. (eds) (1990) *Sustaining Earth*, Macmillan, London.

Ansoff, H. I. (1984) *Implementing Strategic Management*, Prentice-Hall, Englewood Cliffs.

Anthony, R. N. (1965) *Planning and Control Systems*, Harvard Business School Press, Boston.

Arnold, M., Long, F. and Choi, P. (1991) *Marketing and Ecology*, Management Institute for Environment and Business, Washington DC.

Ashby, E. (1978) *Reconciling Man with the Environment*, Oxford University Press, Oxford.

Attfield, R. (1983) *The Ethics of Environmental Concern*, Blackwell, Oxford.

Ausubel, J. H. and Sladovich, H. E. (eds) (1989) *Technology and Environment*, National Academy Press, Washington DC.

Baark, E. and Svedin, U. (eds) (1988) *Man, Nature and Technology*, Macmillan, London.

Badaracco, J. L. (1991) *The Knowledge Link*, Harvard Business School Press, Cambridge.

Baker, M. J. (1974) *Marketing*, Macmillan, London.

Ball, S. and Bell, S. (1992) *Environmental Law*, Blackstone Press, London.

Balling, R. C. (1992) *The Heated Debate*, Pacific Research Institute for Public Policy, San Francisco.

Banks, R. (ed) (1989) *Costing the Earth*, Shepherd Walwyn, London.

Banister, D. and Button, K. (eds) (1993) *Transport, the Environment and Sustainable Development*, Spon, London.

Barbier, E. G. (1989) *Economics, Natural Resource Scarcity and Development*, Earthscan, London.

Barbier, E. G., Burgess, J., Swanson, T. and Pearce, D. W. (1990) *Elephants, Economics and Ivory*, Earthscan, London.

Barbour, I. G. (1980) *Technology, Environment and Human Values*, Praeger, New York.

Barde, J. P. and Pearce, D. W. (eds) (1991) *Valuing the Environment*, Earthscan, London.

Barker, T. (ed.) (1991) *Green Futures for Economic Growth*, Cambridge Econometrics, Cambridge.

Barney, G. O. (1980) *The Global 2000 Report to the President of the US*, Pergamon, Oxford.

Barrett, S. (1991) 'Environmental regulation for competitive advantage', *Business Strategy Review*, pp. 1–15.

Barrett, B. F. D. and Therival, R. (1992) *Environmental Policy and Impact Assessment in Japan*, Routledge, London.

Barry, N. (1991) *The Morality of Business Enterprise*, Aberdeen University Press, Aberdeen.

Bartlett, C. and Ghoshal, S. (1989) *Managing Across Borders*, Hutchinson, London.

Bateson, G. (1988) *Mind and Nature*, Bantam, New York.

Baumol, W. J. and Oates, W. E. (1979) *Economics, Environmental Policy and the Quality of Life*, Prentice-Hall, Englewood Cliffs.

Baumol, W. J. and Oates, W. E. (1975) *The Theory of Environmental Policy*, Prentice-Hall, Englewood Cliffs.

Beaumont, J. R. (1991) *An Introduction to Market Analysis*, Geo Books, Norwich.

Beaumont, J. R. (1992) 'Managing our environment', Futures, April, pp. 187–205.

Beaumont, J. R. (1992) 'Towards the 21st Century – the way forward', *Interspectives*, 11, pp. 25–28.

Beaumont, J. R. (1993) 'The greening of the car industry', *Environment and Planning A*, 25(7), pp. 909–922.

Beaumont, J. R. and Sutherland, E. (1992) *Information Resources Management*, Butterworth-Heinemann, Oxford.

Beaumont, J. R. and Keys, P. (1982) *Future Cities*, John Wiley and Sons, London.

Beckerman, W. (1974) *In Defence of Economic Growth*, Cape, London.

Beckerman, W. (1990) *Pricing for Pollution*, Institute for Economic Affairs, London.

Beer, M., Spector, B., Lawrence, P., Mills, Q. and Walton, R. (1984) *Managing Human Assets*, Free Press, New York.

Beesley, M. and Evans, T. (1978) *Corporate Social Responsibility*, Croom Helm, London.

Belasco, J. A. (1990) *Teaching the Elephant to Dance*, Hutchison Business Books, London.

Bell, D. (1973) *The Coming of Post-Industrial Society*, Penguin, Harmondsworth.

Benedikt, M. (1991) *Cyberspace*, MIT Press, Cambridge.

Bennet, S. J. (1991) *Ecopreneuring*, John Wiley and Sons, New York.

Berkes, F. (ed.) (1991) *Common Property Resources*, Belhaven, London.

Bernstam, M. (1991) *The Wealth of Nations and the Environment*, Institute for Economic Affairs, London.

Bernstein, D. (1992) *In the Company of Green*, ISBA, London.

Berry, T. (1988) *The Dream of the Earth*, Sierra Club Books, San Francisco.

Berry, W. (1975) *A Continuous Harmony*, Harcourt Brace, New York.

Bhalla, A. S. (ed.) (1992) *Environment, Employment and Development*, International Labour Office, Geneva.

Birch, J. (1976) *Confronting the Future*, Penguin, Harmondsworth.

Birnie, P. and Boyle, A. (1993) *International Law and the Environment*, Clarendon Press, Oxford.

Black, J. (1970) *The Dominion of Man*, Edinburgh University Press, Edinburgh.

Blaikie, P. (1991) *The Political Economy of Soil Erosion in Developing Countries*, Longman, London.

Blumenfeld, K.; Earle, R. and Annighofer, F. (1992) 'Environmental performance and business strategy', *Prism*, (fourth quarter), pp. 65–81.

Bohi, D. R. (1981) *Analysing Demand Behaviour*, Johns Hopkins University Press, Baltimore.

Bojo, J., Maler, K-G. and Unemo, L. (1990) *Environment and Development*, Kluwer, Dordrecht.

Bolin, B., Doss, B. R., Jager, J. and Warrwick, R. A. (1989) *The Greenhouse Effect, Climatic Change and Ecosystems*, John Wiley and Sons, Chichester.

Bookchin, M. (1982) *The Ecology of Freedom*, Palo Alto, California.

Boserup, E. (1981) *Population and Technology*, Blackwell, Oxford.

Botkin, D. B., Caswell, M. F., Estes, J. E. and Orio, A. A. (1989) *Changing the Global Environment*, Academic Press, Boston.

Boyle, S. and Ardill, J. (1989) *The Greenhouse Effect*, Hodder and Stoughton, London.

Braden, J. and Kolstad, C. (1991) *Measuring the Demand for Environmental Quality*, North Holland-Elsevier, Amsterdam.

Bramwell, A. (1989) *Ecology in the Twentieth Century*, Yale University Press, New Haven.

Brandt, W. (1980) *North-South*, Pan, London.

Breheny, M. J. (ed) (1992) *Sustainable Development and Urban Form*, Pion, London.

Brennan, A. (1988) *Thinking about Nature*, Routledge, London.

British Institute of Management (1991) *Managing the Environment*, BIM, Corby.

Brown, L. R. (1981) *Building a Sustainable Society*, Norton, New York.

Brown, L. R., Chandler, W., Dunning, A., Flavin, C., Heise, L., Jacobson,L., Postel, S., Shen, C., Starke, L. and Wolfe, E. (1988) *State of the World 1988*, Norton, New York.

Brundtland, G. (1987) *Our Common Future*, Oxford University Press, Oxford.

Buchholz, R. A., Marcus, A. A. and Post, J. E. (1992) *Managing Environmental Issues*, Prentice-Hall, Englewood Cliffs.

Burke, T.; Robins, N. and Trisoglio, A. (eds) (1991) *Environmental Strategy Europe 1991*, Camden Publishing, London.

Business International (1990) *Managing the Environment*, The Economist Group, London.

Business in the Environment (1992) *A Measure of Commitment*, BIE (with KPMG Peat Marwick), London.

Button, J. (1989) *How to be Green*, Century Hutchinson, London.

Buzzell, R. D. and Gale, B. T. (1987) *The PIMS Principles*, Free Press, New York.

Cairncross, F. (1991) *Costing the Earth*, Economist Books, London.

Caldwell, L. K. (1992) *Between Two Worlds*, Cambridge University Press, Cambridge.

Callenbach, E., Capra, E. and Marburg, S. (1990) *The Elmwood Guide to Eco-Auditing and Ecologically Conscious Management*, Elmwood Institute, Berkeley.

Cannon, T. (1992) *Corporate Responsibility*, Financial Times/Pitman, London.

Capel, J. (1989) *Green Book of British Environmental Investment Opportunities*, James Capel, London.

Capra, F. (1983) *The Turning Point*, Fontana, London.

Carlzon, J. (1987) *Moments of Truth*, Harper and Row, New York.

Carson, P. and Moulden, J. (1991) *Green is Gold*, Harper Business, Toronto.

Carson, R. (1956) *The Sea Around Us*, Penguin, Harmondsworth.

Carson, R. (1963) *Silent Spring*, Penguin, Harmondsworth.

Central Computer and Telecommunications Agency (1990) *Managing Information as a Resource*, CCTA, HMSO, London.

CERES (1990) *The 1990 CERES Guide to the Valdez Principles*, CERES, Boston.

Chandler, A. J. (1962) *Strategy and Structure*, MIT Press, Cambridge.

Chandler, A. J. (1990) *Scale and Scope*, Harvard University Press and Belknap, Cambridge.

Charter, M. (ed.) (1992) *Greener Marketing*, Greenleaf Publishing, Sheffield.

Chiras, D. D. (1991) *Environmental Science*, Benjamin/Cummings, Redwood City.

Christopher, M., McDonald, M. and Wills, G. (1980) *Introducing Marketing*, Pan, London.

Cline, W. (1991) *Estimating the Benefits of Greenhouse Warming Abatement*, OECD, Paris.

Clutterbuck, D. and Snow, D. (1990) *Working with the Community*, Weidenfeld and Nicolson, London.

Coddington, W. and Florian, P. (1993) *Environmental Marketing*, McGraw-Hill, London.

Coker, A. and Richards, C. (eds) (1992) *Valuing the Environment*, Belhaven Press, London.

Collard, D., Pearce, D. W. and Ulph, D. (eds) (1988) *Economics, Growth and Sustainable Environments*, Macmillan, London.

Commoner, B. (1972) *The Closing Circle*, Alfred Knopf, New York.

Commoner, B. (1990) *Making Peace with the Planet*, Pantheon, New York.

Confederation of British Industry (1990) *Waking Up to a Better Environment*, CBI and PA Consulting, London.

Conroy, C. and Litvinoff, M. (eds) (1988) *The Greening of Aid*, Earthscan, London.

Consumer Association (1990) 'Green labelling', *Which?*, January, London.

Conway, G. R. and Barbier, E. B. (1990) *After the Green Revolution*, Earthscan, London.

Conway, G. R. and Barbier, E. B. (1990) *Sustainable Agriculture for Development*, Earthscan, London.

Coopers & Lybrand (1985) *A Challenge to Complacency*, Manpower Services Commission, London.

Coopers & Lybrand Deloitte (1990) *Environment and the Finance Function*, Coopers & Lybrand Deloitte, London.

Coopers & Lybrand Deloitte (1991) *Your Business and the Environment*, Coopers & Lybrand Deloitte, London.

Costanza, R. (ed.) (1991) *Ecological Economics*, Columbia University Press, New York.

Cotgrove, S. (1982) *Catastrophe or Cornucopia*, John Wiley and Sons, New York.

Council on Economic Priorities (1989) *Shopping for a Better World*, CEP, New York.

Council on Economic Priorities (1991) *The Better World Investment Guide*, CEP, New York.

Crawford, R. (1991) *In the Era of Human Capital*, Harper Business, New York.

Crossland, A. (1971) *A Sound Democratic Britain*, Fabian Society, London.

Crozier, M. (1990) *L'Organisation à l'Ecoute*, Les Editions d'Organisation, Paris.

Curran, J. and Seaton, J. (1988) *Power without Responsibility*, Routledge, London.

Cutter, S. L. (1993) *Living with Risk*, Edward Arnold, London.

Daly, H. E. (1987) 'The economic growth debate', *Journal of Environmental Economics and Management*, 14 (4), pp. 323–336.

Daly, H. E. (1977) *Steady State Economics*, W. H. Freeman & Company, San Francisco.

Daly, H. E. and Cobb, J. E. (1990) *For the Common Good*, Greenprint, London.

Daniels, N. (ed) (1975) *Reading Rawls*, Blackwell, Oxford.

Dauncey, G. (1988) *After the Crash*, Greenprint, Basingstoke.

David Bellamy Associates (1991) *Industry Goes Green*, Durham.

Davidow, W. H. and Uttal, B. (1989) *Total Customer Service*, Harper and Row, New York.

Davidson, J. D. and Rees-Mogg, W. (1992) *The Great Reckoning*, Sidgwick, London.

Davies, R. L. and Reynolds, J. (1988) *The Development of Teleshopping and Teleservices*, Longman, Harlow.

Davis, J. (1991) *Greening Business*, Blackwell, Oxford.

Davis, K. and Blomstrom, R. L. (1975) *Business and Society*, McGraw-Hill, New York.

de Bono, E. (1992) *Sur/Petition*, Harper Business, New York.

Deal, T. E. and Kennedy, A. (1988) *Corporate Culture*, Penguin, Harmondsworth.

Department for Education and the Welsh Office (1993) *Environmental Responsibility*, HMSO, London.

Department of the Environment (1989) *Clean Technology*, DoE, London.

Department of the Environment (1989) *Environment in Trust*, DoE, London.

Department of the Environment (1989) *Sustaining Our Common Future*, DoE, London.

Department of the Environment (1990) *This Common Inheritance*, HMSO, London.

Department of Trade and Industry (1989) *Your Business and the Environment*, DTI, London.

Department of Trade and Industry (1990) *Cutting Your Losses*, DTI, London.

Dobson, A. (1990) *Green Political Thought*, Unwin Hyman, London.

Dorfman, R. and Dorfman, N. S. (eds) (1977) *Economics of the Environment*, Norton, New York.

Dower, N. (ed) (1989) *Ethics and Environmental Philosophy*, Avebury, Aldershot.

Dreze, J. and Sen, A. (1989) *Hunger and Public Action*, Clarendon Press, Oxford.

Drucker, P. F. (1954) *The Practice of Management*, Heinemann, New York.

Drucker, P. F. (1989) *The New Realities*, Harper & Row, New York.

Drucker, P. F. (1991) 'The new productivity challenge', *Harvard Business Review*, 69 (6) pp. 69–79.

Drucker, P. F. (1992) *For the Future*, Dutton, New York.

Drucker, P. F. (1993) *The Post-Capitalist Society*, Butterworth-Heinemann, Oxford.

Drucker, P. F. (1993) *The Ecological Vision*, Transaction, New York.

Durning, A. (1992) *How Much is Enough?*, Earthscan, London.

Eckersley, R. (1992) *Environmentalism and Political Theory*, UCL Press, London.

Eckholm, E. P. (1976) *Losing Ground*, Norton, New York.

Ehrlich, A. H. and Ehrlich, P. R. (1987) *Earth*, Franklin Watts, New York.

Ehrlich, P. R. (1968) *The Population Bomb*, Ballantine, New York.

Ehrlich, P. R. (1986) *The Machinery of Nature*, Simon and Schuster, New York.

Ehrlich, P. R. and Ehrlich, A. H. (1991) *The Population Explosion*, Arrow, London.

Ehrlich, P. R., Erhlich, A. H. and Holdren, J. P. (1977) *Ecoscience*, W H Freeman, San Francisco.

Ehrlich, P. R. and Holdren, J. P. (1974) 'Impacts of population growth', *Science*, 171, pp. 1212–1217.

Ehrlich, P. R. and Roughgarden, R. (1987) *The Science of Ecology*, Macmillan, New York.

Ekins, P. (ed.) (1986) *The Living Economy,* Routledge and Kegan Paul, London.

Ekins, P., Hillman, M. and Hutchinson, R. (1992) *Wealth Beyond Measure,* Gaia Press, London.

Elkington, J. (1990) *The Environmental Audit,* Sustainability, London.

Elkington, J. and Hailes, J. (1988) *The Green Consumer Guide,* Victor Gollancz, London.

Elkington, J. and Hailes, J. (1989) *The Green Consumer's Supermarket Shopping Guide,* Victor Gollancz, London.

Elkington, J. and Hailes, J. (1991) *The Green Business Guide,* Victor Gollancz, London.

Elkington, J. and Burke, T. (1989) *The Green Capitalists,* Victor Gollancz, London

Etzioni, A. (1988) *The Moral Dimension,* Free Press, New York.

Eyre, S. R. (1979) *The Real Wealth of Nations,* Edward Arnold, London.

Eysenck, H. J. (1973) *The Inequality of Man,* Temple Smith, London.

Fagan, B. M. (1990) *The Journey from Eden,* Thames and Hudson, New York.

Fardoust, S. and Dhareshwar, A. (1990) *A Long-term Outlook for the World Economy,* World Bank, Washington DC.

Farman, J. (1987) 'What hope for the ozone layer now?', *New Scientist,* 116(1586), pp. 50–54.

Faulkner, J. H. (1992) 'Foreward', Business Strategy and the Environment, 1(1), pp. i–iii.

Feshbach, M. and Friendly, A. (1992) *Ecocide in the USSR,* Basic Books, New York.

Firor, J. (1990) *The Changing Atmosphere,* Yale University Press, New Haven.

Forrester, S. (1990) *Business and Environmental Groups,* Directory of Social Change, London.

Foster, C. D. (1992) *Privatisation, Public Ownership and the Regulation of Natural Monopoly,* Blackwell, Oxford.

Fowler, C. and Mooney, P. (1990) *Shattering,* University of Arizona Press, Tuscon.

Freeman, A. M. (1979) *The Benefits of Environmental Improvement,* Johns Hopkins University Press, Baltimore.

Freeman, R. E. and Gilbert, D. R. (1980) *Corporate Strategy and the Search for Ethics,* Prentice-Hall, Englewood Cliffs.

French, P. (1984) *Corporations and Corporate Responsibility,* Columbia University Press, New York.

Friedman, F. B. (1991) *A Practical Guide to Environmental Management,* Environmental Law Institute, Washington DC.

Friedman, M. (1962) *Capitalism and Freedom,* University of Chicago Press, Chicago.

Fukuyama, F. (1992) *The End of History and the Last Man*, Hamilton, New York.

Galbraith, J. K. (1974) *The New Industrial State*, Penguin, Harmondsworth.

Galbraith, J. K. (1958) *The Affluent Society*, Hamish Hamilton, London.

Gallup, G. H. International Institute (1992) *The Health of the Planet Survey*, Gallup, Princeton.

Garvin, D. (1988) *Quality Management*, Free Press, New York.

Georgescu-Roegen, N. (1971) *The Entropy Law and the Economic Process*, Harvard University Press, Cambridge.

Glacken, C. J. (1970) *Traces on the Rhodian Shore*, University of California Press, Berkeley.

Goldsmith, E. (1972) *A Blueprint for Survival*, Penguin, Harmondsworth.

Goldsmith, E. (1992) *The Way*, Rider, London.

Goldsmith, E., Allen, R., Allaby, M., Davoll, J. and Lawrence, S. (1972) *A Blueprint for Survival*, The Ecologist.

Goldsmith, E. and Hildyard, N. (eds) (1988) *The Earth Report*, Mitchell Beazley, London.

Goldsmith, E. and Hildyard, N. (eds) (1990) *The Earth Report 2*, Mitchell Beazley, London.

Goldsmith, E. and Hildyard, N. (eds) (1992) *The Earth Report 3*, Mitchell Beazley, London.

Goldsmith, W. and Clutterbuck, D. (1984) *The Winning Streak*, Weidenfeld and Nicolson, London.

Goodin, R. E. (1992) *Green Political Theory*, Polity Press, Cambridge.

Goold, M. and Campbell, A. (1987) *Strategies and Styles*, Blackwell, Oxford.

Gordon, A. and Suzuki, D. (1991) *It's a Matter of Survival*, Harvard University Press, Cambridge.

Gore, A. (1992) *Earth in the Balance*, Earthscan, London.

Gorry, G. A. and Scott Morton, M. S. (1989) 'A framework for Management Information Systems', *Sloan Management Review*, 30 (3) pp. 49–61 (reprinted from 1971).

Gosovic, B. (1992) *The Quest for World Environmental Cooperation*, Routledge, Chapman and Hall, Andover.

Goudie, A. S. (1990) *The Human Impact on the Natural Environment*, Blackwell, Oxford.

Grainger, A. (1990) *The Threatening Desert*, Earthscan Publications, London.

Grant, R. M. (1991) *Contemporary Strategy Analysis*, Blackwell, Oxford.

Gray, R. H. (1990) *The Greening of Accountancy*, ACCA, London.

Gray, R. H., Owen, D. and Maunders, K. (1987) *Corporate Social Reporting*, Prentice-Hall, Englewood Cliffs.

Gray, P., King, W. R., McLean, E. R. and Watson, H. J. (1989) *Management of Information Systems*, Dryden Press, Chicago.

Green Party (1986) *Manifesto for a Sustainable Society*, Green Party, London.

Greenley, G. E. (1989) *Strategic Management*, Prentice-Hall, Englewood Cliffs.

Grenon, M. and Batisse, M. (eds) (1991) *Futures for the Mediterranean Basin*, Oxford University Press, Oxford.

Gribben, J. (1988) *The Hole in the Sky*, Bantam, New York.

Gupta, A. (1988) *Ecology and Development in the Third World*, Routledge, London.

Haag, E. van den (1976) 'Economic is not enough - notes on the anticapitalist spirit', *The Public Interest*, (Fall), p. 109.

Haavind, R. (1992) *The Road to the Baldrige Award*, Butterworth-Heinemann, Boston.

Haigh, N. and Baldcock, D. (1989) *Environmental Policy and 1992*, Institute for European Environmental Policy, London.

Haigh, N. (1989) *EEC Environmental Policy and Britain* Longman, Harlow.

Haigh, N. (1992) *Manual of Environmental Policy*, Longman, London.

Halal, W. E. (1986) *The New Capitalism*, John Wiley and Sons, New York.

Halberstam, D. (1991) *The Next Century*, Morrow, New York.

Hancock, G. (1989) *Lords of Poverty*, Macmillan, London.

Handy, C. (1989) *The Age of Unreason*, Business Books, London.

Hardin, G. (1968) 'The tragedy of the commons', *Science*, 162, pp. 1243–1248.

Hardoy, J. E., Cairncross, S. and Satterthwaite, D. (eds) (1990) *The Poor Die Young*, Earthscan, London.

Hassan, H. M. and Hutchinson, C. (eds) (1992) *Natural Resource and Environmental Information for Decisionmaking*, World Bank, Washington DC.

Hax, A. C. and Majluf, N. S. (1984) *Strategic Management*, Prentice-Hall, Englewood Cliffs.

Hayek, F. A. (1960) *The Constitution of Liberty*, University of Chicago Press, Chicago.

Hayek, F. A. (1976) *The Mirage of Social Justice*, Routledge and Kegan Paul, London.

Hazarika, S. (1987) *Bhopal*, Penguin, New Delhi.

Hecht, S. and Cockburn, A. (1989) *The Fate of the Forest*, New Left Books, London.

Henley Centre (1991) *Young Eyes*, Henley Centre (and BT), London.

Hennison, K. and Kinnear, T. (1976) *Ecological Marketing*, Prentice-Hall, Englewood Cliffs.

Her Majesty's Stationery Office (1990) *Developments in Biotechnology*, HMSO, London.

Hick, J. (1946) *Value and Capital*, Oxford University Press, Oxford.

Hill, T. (1989) *Manufacturing Strategy*, Macmillan, London.

Hirsil, F. (1977) *Social Constraints to Growth*, Routledge and Kegan Paul, London.

Hoffman, W. M., Frederick, R. and Petry, E. S. (1990) *The Corporation, Ethics and the Environment*, Quorum Books, Westport.

Holmberg, J., Bass, S. and Timberlake, J. (1991) *Defending the Future*, Earthscan, London.

Horta, K. (1991) 'The last big rush for the green gold', *The Ecologist*, 21(3), pp. 142–147.

Houghton, J. T., Callender, B. and Varney, S. (eds) (1992) *Climate Change 1992*, Cambridge University Press, Cambridge.

Houghton, J. T., Jenkins, G. J. Ephraums, J. J. (eds) (1990) *Climate Change*, Cambridge University Press, Cambridge.

Hunt, J. (1991) 'Heseltine attacks business on environment', *Financial Times*, 27th September.

Hurrell, A. and Kingsbury, B. (eds) (1992) *The International Politics of the Environment*, Oxford University Press, Oxford.

Hussey, D. E. (ed.) (1990) *International Review of Strategic Management*, John Wiley and Sons, London.

Hutchinson, C. (1991) *Business and the Environmental Challenge*, The Conservation Trust, Reading.

Hyde, W. F. and Newman, D. H. (1991) *Forest Economics and Policy Analysis*, The World Bank, Washington DC.

Infoplan (1992) *Environmentalism*, Infoplan, Tokyo.

Institute of Directors (1992) *Stewards of the Earth*, IOD, London.

Institute of Business Ethics (1990) *Ethics, Environment and the Company*, IBE, London.

International Labour Organization (1992) *Environment and the World of Work*, ILO, Geneva.

International Union for the Conservation of Nature (1980) *World Conservation Strategy*, IUCN, Gland.

International Union for the Conservation of Nature (1991) *Caring for the Earth*, Earthscan, London.

International Chamber of Commerce (1989) *Environmental Auditing*, ICC, Paris.

Irvine, S. and Ponton, A. (1988) *A Green Manifesto*, Optima, London.

Irvine, S. (1990) 'No growth in a finite world', *New Statesman and Society*, 3 (128), pp. 16–18.

Jack, A. (1992) 'Style without substance', *Financial Times*, 16th September.

Jacks, G. V. and Whyte, R. O. (1939) *The Rape of the Earth*, Faber and Faber, London.

Jackson, B. (1990) *Poverty and the Planet*, Penguin, Harmondsworth.

Jacobs, M. (1991) *The Green Economy*, Pluto, London.

Jacobsen, J. (1988) *Environmental Refugees*, Worldwatch Paper 86, Worldwatch Institute, Washington DC.

James, P. (1992) 'The corporate response', in M. Charter (ed.) *Greener Marketing*, Greenleaf, Sheffield, pp. 111–137.

Johansson, P-O. (1987) *The Economic Theory and Measurement of Environmental Benefits*, Cambridge University Press, Cambridge.

Johnson, H. T. and Kaplan, R. S. (1987) *The Rise and Fall of Management Accounting*, Harvard Business School Press, Cambridge.

Johnson, L. E. (1991) *A Morally Deep World*, Cambridge University Press, Cambridge.

Johnson, H. T. (1992) *Relevance Regained*, Free Press, New York.

Johnson, P., McKay, S. and Smith, S. (1990) *The Distributional Consequences of Environmental Taxes*, Institute for Fiscal Studies, London.

Johnston, R. J. (1991) *A Question of Place*, Blackwell, Oxford.

Joly, C. (1992) 'Green funds, or just greedy?', in D. Koechlin and K. Muller (eds) *Green Business Opportunities*, Pitman, London, pp. 131–153.

Jones, T. M. (1980) 'Corporate social responsibility revisited, redefined', *California Management Review*, 22(2), pp. 59–67.

Jones, R. R. and Wigley, T. (eds) (1989) *Ozone Depletion*, John Wiley, New York.

Joseph, L. E. (1990) *Gaia*, St. Martin's Press, New York.

Juran, J. M. (1992) *Juran on Quality by Design*, Free Press, New York.

Kanter, R. M. (1983) *The Change Masters*, Geore Allen and Unwin, London.

Kanter, R. M. (1989) *When Giants Learn to Dance*, Simon and Schuster, New York.

Kanter, R. M., Stein, B. A. and Jick, J. D. (1992) *The Challenge of Organisational Change*, Free Press, New York.

Keen, P. G. W. (1991) *Shaping the Future*, Harvard Business School Press, Boston.

Kemp, D. D. (1990) *Global Environmental Issues*, Routledge, London.

Kemp, R. (1992) *The Politics of Radioactive Waste Disposal*, Manchester University Press, Manchester.

Kennedy, P. (1993) *Preparing for the Twenty-First Century*, Harper Collins, London.

Keynes, J. M. (1933) 'National self-sufficiency', Yale Law Review, 22, pp. 755–763.

King, A. and Schneider, B. (1991) *The First Global Revolution*, Simon and Schuster, London.

Kirzner, I. (1989) *Discovery, Capitalism and Distributive Justice*, Blackwell, Oxford.

Klein, T. A. (1977) *The Social Costs and Benefits of Business*, Prentice-Hall, Englewood Cliffs.

Kneese, A., Ayres, R. U. and d'Arge, R. (1970) *Economics and the Environment*, Johns Hopkins University Press, Baltimore.

Koechlin, D. and Müller, K. (eds) (1992) *Green Business Opportunities*, Pitman, London.

Kotler, P. (1988) *Marketing Management*, Prentice-Hall, Englewood Cliffs.

Kotter, J. P. and Heskett, J. L. (1992) *Corporate Culture and Performance*, Free Press, New York.

Kritz, M. M. (1990) 'Climate Change and Migration Adaptations', Working Paper 2.16, Population and Development Program, Cornell University, Ithaca.

Kuhn, J. W. and Shriver, D. W. (1991) *Beyond Success*, Oxford University Press, New York.

Kula, E. (1992) *Economics of Natural Resources and the Environment*, Chapman and Hall, Andover.

Labour Party (1990) *An Earthly Chance*, Labour Party, London.

Lapp, F. M. and Schurman, R. (1989) *Taking Population Seriously*, Earthscan, London.

Lascelles, D. (1992) 'Hot air in Rio sows green seed', *Financial Times*, 29th July, p. 14.

LeBlanc, C. (1991) *The Nature of Growth*, National Andubon Society, Washington DC.

Ledgerwood, G., Street, E. and Therivel, R. (1992) *The Environmental Audit and Business Strategy*, Pitman, London.

Leggett, J. (ed.) (1990) *Global Warming*, Oxford University Press, Oxford.

Lessem, R. (1991) *Total Quality Learning*, Blackwell, Oxford.

Levitt, T. (1987) 'The mixed metrics of greed', *Harvard Business Review*, 65(6), (editorial), pp. 6–7.

Lipietz, A. (1992) *Towards a New Economic Order*, Polity Press, London.

Lloyd, T. (1990) *The 'Nice' Company*, Bloomsbury, London.

Lockyer, K., Muhlemann, A. and Oakland, J. (1991) *Production and Operations Management*, Pitman, London.

Lorenz, C. (1987) *The Design Dimension*, Blackwell, Oxford.

Lovelock, J. E. (1979) *Gaia*, Oxford University Press, Oxford.

Lovelock, J. E. (1989) *The Ages of Gaia*, Oxford University Press, Oxford.

Lovelock, J. E. (1990) 'Hands up for the Gaia hypothesis', *Nature*, 344, pp. 100–102.

Luthans, F., Hodgetts, R. M. and Thompson, K. R. (1984) *Social Issues in Business*, Macmillan, New York.

Lutz, M. and Lux, K. (1988) *Humanistic Economics*, Bootstrap Press, New York.

MacNeil, J., Winsemius, P. and Yakushiji, Y. (1991) *Beyond Interdependence*, Oxford University Press, New York.

Malecki, E. J. (1991) *Technology and Economic Development*, Longman, Harlow.

Mangham, I. L. and Pye, A. (1991) *The Doing of Managing*, Blackwell, Oxford.

Mannion, A. M. (1991) *Global Environmental Change*, Longman, Harlow.

Marquand, D. (1988) *The Unprincipled Society*, Jonathan Cape, London.

Marshall, P. (1992) *Nature's Web*, Simon and Schuster, New York.

Martin, C. (1991) *The Rainforests of West Africa*, Birkhauser Verlag, Basle.

Marx, J. E. (ed.) (1989) *A Revolution in Biotechnology*, Cambridge University Press, Cambridge.

McCarta, R. (1990) *The Gaia Atlas of Future Worlds*, Gaia Books, London.

McKibben, B. (1990) *The End of Nature*, Penguin, Harmondsworth.

McKinnon, A. C. (1989) *Physical Distribution Systems*, Routledge, London.

Meadows, D. H., Meadows, D. L. and Randers, J. (1992) *Beyond the Limits*, Chelsea Green, Vermont.

Meadows, D. H., Meadows, D. L., Randers, J. and Behrens, W. W. (1972) *The Limits to Growth*, Universe Books, New York.

Medawar, C. (1978) *The Social Audit Consumer Handbook*, Macmillan, London.

Mills, J. S. (1891) *Principles of Political Economy*, Routledge, London.

Mills, D. Q. (1991) *Rebirth of the Corporation*, John Wiley and Sons, New York.

Milsted, D. (1990) *The Green Bluffer's Guide*, Ravette Books, Horsham.

Mintzberg, H. (1990) 'Strategy formation: schools of thought', in J. W. Fredrickson (ed) *Perspectives on Strategic Management*, Harper and Row, New York, pp. 105–236.

Mintzberg, H. (1989) *Mintzberg on Management*, Free Press, New York.

Mintzer, I. M. (ed.) (1992) *Confronting Climate Change*, Cambridge University Press, Cambridge.

Mishan, E. (1967) *The Costs of Economic Growth*, Penguin, Harmondsworth.

Moran, A., Chisholm, A. and Porter, M. (eds) (1991) *Markets, Resources and the Environment*, Allen and Unwin, Sydney.

Morris, M. D. (1979) *Measuring the Condition of the World's Poor*, Pergamon, Oxford.

Mungall, C. and McLaren, D. J. (eds) (1990) *Planet under Stress*, Oxford University Press, Oxford.

Myers, N. (1979) *The Sinking Ark*, Pergamon, New York.

Myers, N. (1983) *A Wealth of Wild Species*, Westview, Boulder.

Myers, N. (1984) *The Primary Source*, Norton, New York.

Myers, N. (ed.) (1985) *The Gaia Atlas of Planet Management*, Pan, London.

Myers, N. (1990) *The Gaia Atlas of Future Worlds*, Gaia Books, London.

Myrdal, G. (1957) *Economic Theory and Underdeveloped Regions*, Duckworth, London.

Nader, R. (ed.) (1973) *The Consumer and Corporate Accountability*, Harcourt Brace Jovanovich, New York.

Nader, R., Green, M. and Seligman, J. (1976) *Taming the Corporate Giant*, Norton, New York.

Naisbitt, J. and Aburdene, P. (1990) *Megatrends 2000*, Sidgwick & Jackson, London.

Neimark, M. (ed.) (1988) *Advances in Public Interest Accounting*, JAI Press, Greenwich.

Netherlands Ministry of Housing, Physical Planning and the Environment (1989) *National Environmental Policy Plan of the Netherlands*, Amsterdam.

Niebuhr, H. R. (1963) *The Responsible Self*, Harper & Row, New York.

Nilsson, A. (1992) *Greenhouse Earth*, John Wiley and Sons, Chichester.

Nisbet, E. G. (1992) *Leaving Eden*, Cambridge University Press, Cambridge.

Nordhaus, W. D. (1990) *To Slow or Not to Slow*, Yale University Press, New Haven.

Nordhaus, W. D. (1991) 'A sketch of the economics of the greenhouse effect', *American Economic Review*, 81(2), pp. 146–150.

Norgaard, R. (1992) *Sustainability and the Economics of Assuring Assets for Future Generations*, The World Bank, Washington DC.

Nozick, R. (1974) *Anarchy, State and Utopia*, Basic Books, New York.

O'Riordan, T. (1981) *Environmentalism*, Pion, London.

O'Riordan, T. (1989) 'The challenge for environmentalism', in R. Peet and N. J. Thrift (eds) *New Models in Geography*, Unwin, London.

Oasis (1989) *Management of Marketing Information*, Institute of Marketing, London.

Odell, P. R. (1986) *Oil and World Power*, Penguin, Harmondsworth.

Odum, H. T. (1983) *Systems Ecology*, John Wiley and Sons, New York.

Odum, P. E. (1975) *Ecology*, Holt, Rinehart and Winston, New York.

Ohmae, K. (1990) *The Borderless World*, Collins, Glasgow.

Okun, A. (1975) *Equality and Efficiency*, The Brookings Institute, Washington DC.

Opschoor, J. and Vos, H. (eds) (1989) *The Application of Economic Instruments for Environmental Protection in OECD Member Countries*, OECD, Paris.

Organization for Economic Cooperation and Development (1975) *The Polluter Pays Principle*, OECD, Paris.

Organization for Economic Cooperation and Development (1985) *The Macroeconomic Impact of Environmental Expenditure*, OECD, Paris.

Organization for Economic Cooperation and Development (1989) *The Application of Economic Instruments for Environmental Protection*, OECD, Paris.

Organization for Economic Cooperation and Development (1989) *Environmental Policy Benefits*, OECD, Paris.

Organization for Economic Cooperation and Development (1989) *Environmental Management in Developing Countries*, OECD, Paris.

Organization for Economic Cooperation and Development (1991) *The State of the Environment*, OECD, Paris.

Ornstein, R. and Ehrlich, P. R. (1989) *New World/New Mind*, Doubleday, New York.

Ostrom, E. (1990) *Governing the Commons*, Cambridge University Press, Cambridge.

Owen, D. (ed.) (1992) *Green Reporting*, Chapman and Hall, London.

Paehlke, R. C. (1989) *Environmentalism and the Future of Progressive Politics*, Yale Unversity Press, New Haven.

Parkinson, L. K. and Parkinson, S. T. (1987) *Using the Microcomputer in Marketing*, McGraw-Hill, New York.

Parry, M. (1990) *Climate Change and World Agriculture*, Earthscan, London.

Parry, M., Carter, T. R. and Konijn, N. T. (eds) (1988) *The Impact of Climatic Variations on Agriculture*, Dordrecht, Amsterdam.

Passmore, J. (1974) *Man's Responsibity for Nature*, Duckworth, London.

Pearce, D. W. (1989) *Tourist Development*, Longman, London.

Pearce, D. W. (ed.) (1991) *Blueprint 2*, Earthscan, London.

Pearce, D. W. (1993) *Economic Values and the Natural World*, Earthscan, London.

Pearce, D. W. and Atkinson, G. (1992) *Are National Economies Sustainable?*, CSERGE, University College, London.

Pearce, D. W., Bann, C. and Georgiou, C. (1992) *The Social Cost of Fuel Cycles*, HMSO, London.

Pearce, D. W., Barbier, E. B. and Markandya, A. (1990) *Sustainable Development*, Earthscan, London.

Pearce, D. W., Markandya, A. and Barbier, E. B. (1989) *Blueprint for a Green Economy*, Earthscan, London.

Pearce, D. W., Markandya, A. and Barbier, E. B. (1990) 'Environmental sustainability and cost-benefit analysis', *Environment and Planning A*, 22, pp. 1259–1266.

Pearce, D. W. and Turner, R. K. (1990) *The Economics of National Resources and the Environment*, Harvester-Wheatsheaf, Hemel Hempstead.

Pearce, D. W. and Warford, J. (1992) *World Without End*, Oxford University Press, Oxford.

Peattie, K. (1992) *Green Marketing*, Pitman, London.

Pepper, D. (1984) *The Roots of Modern Environmentalism*, Croom Helm, London.

Perry, D. L. (1976) *Social Marketing Strategies*, Goodyear, Pacific Palisades.

Peters, T. J. (1987) *Thriving on Chaos*, Alfred A. Knopf, New York.

Peters, T. J. (1992) *Liberation Management*, Macmillan, London.

Peters, T. J. and Waterman, R. H. (1982) *In Search of Excellence*, Harper and Row, New York.

Pettigrew, A. M. and Whipp, R. (1991) *Managing Change for Competitive Success*, Blackwell, Oxford.

Pirsig, R. M. (1974) *Zen and the Art of Motorcycle Maintenance*, William Morrow, New York.

Plant, C. and Plant, J. (1991) *Green Business*, Green Books, Bideford.

Polanyi, M. (1967) *The Tacit Dimension*, Routledge and Kegan Paul, London.

Poore, D. (1989) *No Timber Without Trees*, Earthscan, London.

Porritt, J. (1984) *Seeing Green*, Basil Blackwell, Oxford.

Porritt, J. (1990) *Where on Earth are We Going?*, BBC, London

Porritt, J. and Winner, D. (1988) *The Coming of the Greens*, Fontana, London.

Porteous, A. (1992) *Dictionary of Environmental Science*, John Wiley And Sons, Chichester.

Porter, M. E. (1980) *Competitive Strategy*, Free Press, New York.

Porter, M. E. (1983) *Cases in Competitive Strategy*, Free Press, New York.

Porter, M. E. (1985) *Competitive Advantage*, Free Press, New York.

Porter, M. E. (1990) *The Competitive Advantage of Nations*, Free Press, New York.

Porter, M. E. (1991) 'America's green strategies', *Scientific American*, April, p. 168.

Prahalad, C. K. and Hamel, G. (1990) 'The core competence of the corporation', Harvard Business Review, 68 (3) pp. 79–91.

Preston, L. (ed.) (1990) *Government Regulation and Business Response*, JAI Press, Greenwich.

Quinn, J. B. (1992) *Intelligent Enterprise*, Free Press, New York.

Ralston, K. and Church, C. (1991) *Working Greener*, Greenprint, London.

Rapp, S. and Collins, T. (1990) *The Great American Marketing Turnaround*, Prentice-Hall, Englewood Cliffs.

Rawls, J. (1971) *A Theory of Justice*, Harvard University Press, Cambridge.

Redcliffe, M. (1984) *Development and the Environmental Crisis*, Methuen, London.

Redcliffe, M. (1987) *Sustainable Development*, Methuen, London.

Rees, J. (1990) *Natural Resources*, Routledge, London.

Reich, C. (1970) *The Greening of America*, Random House, New York.

Reich, R. (1991) *The World of Work*, Alfred A Knopf, New York.

Reid, W. V. and Miller, K. R. (1989) *Keeping Options Alive*, World Resources Institute, Washington DC.

Reidenbach, R. and Robin, D.P. (1989) *Ethics and Profits*, Prentice-Hall, Englewood Cliffs.

Reisman, D. (1980) *Galbraith and Market Capitalism*, Macmillan, London.

Repetto, R. (ed.) (1985) *The Global Possible*, Yale University Press, New Haven.

Repetto, R. (1989) *Wasting Assets*, World Resources Institute, Washington.

Repetto, R. and Gillis, M. (1988) *Public Policies and the Misuse of Forest Resources*, Cambridge University Press, Cambridge.

Rheingold, H. (1991) *Virtual Reality*, Secker and Warburg, New York.

Robertson, J. (1990) *Future Wealth*, Cassell, London.

Robins, N. (1990) *Managing the Environment*, Business International, London.

Robins, N. (1990) *The Quality Route to the Environment*, Business International, London.

Robinson, H. (1981) *Population and Resources*, Macmillan, London.

Robinson, M. (1992) *The Greening of British Party Politics*, Manchester University Press, Manchester.

Robinson, W. and Bolen, E. (1989) *Wildlife Ecology and Management*, Macmillan, New York.

Roddick, A. (1991) *Body and Soul*, Ebury Press, London.

Rogers, K. (1991) 'Who is the greenest of them all?', *Autocar and Motor*, 12 June, pp. 49–55.

Roll, E. (1973) *A History of Economic Thought*, Faber and Faber, London.

Roome, N. (1992) 'Developing environmental management strategies', *Business Strategy and the Environment*, 1(1), pp. 11–24.

Rostow, W. W. (1960) *The Stages of Economic Growth*, Cambridge University Press, Cambridge.

Roszak, T. (1973) *Where the Westland Ends*, Faber and Faber, London.

Ryle, M. (1988) *Ecology and Socialism*, Radius, London.

Salter, J. R. (1992) *Corporate Environmental Responsibility*, Butterworths, London.

Salter, J. R. (1992) *Directors' Guide to Environmental Issues*, Director Books, Hemel Hempstead.

Sand, P. (1990) *Lessons Learned in Global Environmental Governance*, World Resources Institute, Washington DC.

Sandbach, F. (1980) *Environment*, Basil Blackwell, Oxford.

Schaltegger, S. and Sturm, A. (1992) *Eco-controlling*, in D. Koechlin and K. Müller (eds) *Green Business Opportunities*, Pitman, London, pp. 229–240.

Schein, E. (1987) *Process Consulting*, Addison-Wesley, Reading.

Schein, E. (1989) *Organisational Culture and Leadership*, Jossey-Bass, San Francisco.

Scherer, D. and Attig, T. (eds) (1983) *Ethics and the Environment*, Prentice-Hall, Englewood Cliffs.

Schmidheiny, S. (1992) *Changing Course*, MIT Press, Cambridge.

Schnaiberg, A. (1980) *The Environment*, Oxford University Press, Oxford.

Schneider, S. H. and Londer, R. (1984) *The Coevolution of Climate and Life*, Sierra Club Books, San Francisco.

Schneider, S. and Boston, P. (eds) (1991) *Scientists on Gaia*, MIT Press, Boston.

Schneider, S. H. (1989) *Global Warming*, Sierra Club Books, San Francisco.

Schot, J. (1992) 'Credibility and markets as greening forces for the chemical industry', *Business Strategy and the Environment*, 1(1), pp. 35–43.

Schultz, T. W. (1981) *Economics of Population*, Addison Wesley, New York.

Schumacher, E. F. (1973) *Small is Beautiful*, Harper and Row, New York.

Schumacher, E. F. (1977) *A Guide for the Perplexed*, Jonathan Cape, London.

Schumacher, E. F. (1979) *Good Work*, Jonathan Cape, London.

Scott, J. (1976) *The Moral Economy of the Peasant*, Yale University Press, New Haven.

Scott-Morton, M. S. (ed) (1991) *The Corporation of the 1990's*, Oxford University Press, Oxford.

Seabrook, J. (1990) *The Myth of the Market*, Green Books, Bideford.

Seager, J. (ed.) (1990) *The State of the Earth*, Unwin Hyman, London.

Selman, P. (1992) *Environmental Planning*, Paul Chapman, London.

Sen, A. (1987) *Poverty and Famines*, Clarendon Press, Oxford.

Senge, P. (1990) *The Fifth Discipline*, Doubleday, New York.

Sharp, C. and Jennings, T. (1976) *Transport and the Environment*, Leicester University Press, Leicester.

Simmons, I. G. (1989) *Changing the Face of the Earth*, Blackwell, Oxford.

Simon, H. A. (1977) *The New Science of Management Decision-Making*, Prentice-Hall, Englewood Cliffs.

Simon, J. (1981) *The Ultimate Resource*, Martin Robertson, Oxford.

Simpson, A. (1991) *The Greening of Global Investment*, The Economist Publications, London.

Singh, N. (1989) *Economics and the Crisis of Ecology*, Bellew, London.

Slater, J. (1992) *Director's Guide to Environmental Issues*, Director Books, Hemel Hempstead, Hertfordshire.

Smith, A. (1759) *The Theory of Moral Sentiments*, Clarendon Press, Oxford (1976).

Smith, A. (1974) *The Wealth of Nations*, Penguin, Harmondsworth.

Smith, A. (1776) *An Inquiry into the Nature And Causes of the Wealth of Nations*, Dent, London (1910).

Smith, D. (ed.) (1993) *Business and the Environment*, Paul Chapman, London.

Smith, N. C. (1990) *Morality and the Market*, Routledge, London.

Smith, P. and Warr, K. (eds) (1991) *Global Environmental Issues*, Hodder & Stoughton, London.

Snoeyenbos, M., Almeder, R. and Humber, J. (eds) (1983) *Business Ethics*, Prometheus Books, New York.

Soloman, R. and Hansen, K. (1985) *It's Good Business*, Atheneum, New York.

South Commission (1990) *The Challenge to the South*, Oxford University Press, Oxford.

Stalk, G. and Hout, T. M. (1990) *Competing Against Time*, Free Press, New York.

Stapledon, G. (1971) *Human Ecology*, Charles Knight, London.

Stead, W. E. and Stead, J. G. (1992) *Management for a Small Planet*, Sage, London

Steger, U. (1990) *The Greening of the Board Room*, Business and Society Review, Braintree, Maryland.

Steiner, G. A. and Steiner, J. F. (1980) *Business, Government and Society*, Random House, New York.

Stilwell, E. J., Canty, R. C., Kopf, P. W. and Montron, A. M. (1991) *Packaging for the Environment*, Arthur D Little, New York.

Taylor, A. (1992) *Choosing our Future*, Routledge, Chapman and Hall, Andover.

Thomas, K. (1983) *Man and the Natural World*, Allen Lane, London.

Thurow, L. C. (1992) 'Who owns the twenty-first century?', *Sloan Management Review*, 33(3), pp. 5–17.

Thurow, L. C. (1993) *Head to Head*, Nicholas Brealey, London.

Tietenberg, T. (1990) 'Economic instruments for environmental regulation', *Oxford Review of Economic Policy*, 6(1), pp. 17–33.

Tietenberg, T. (1992) *Environmental and Natural Resource Economics*, Harper Collins, New York.

Tobias, M. (1990) *Voice of the Planet*, Bantam, New York.

Tobias, M. (ed.) (1984) *Deep Ecology*, Avant Books, San Diego.

Toffler, A. (1991) *PowerShift*, Bantam, New York.

Tokar, B. (1987) *The Green Alternative*, Miles, New York.

Touche Ross (1990) *Head in the Clouds or Head in the Sand*, Touche Ross, London.

Trenberth, K. (ed.) (1993) *Climate System Modelling*, Cambridge University Press, Cambridge.

Trevor, M. (1988) *Toshiba's New British Company*, Policy Studies Institute, London.

Tricart, J. and Kieweitdejonge, C. J. (1992) *Ecogeography and Rural Management*, Longman, London.

Turner, B. L., Clark, W. C., Kates, R. W., Richards, J. F., Mathews J. T. and Meyer, W. B. (eds) (1990) *The Earth as Transformed by Human Action*, Cambridge University Press, Cambridge.

Turner, R. K. (ed.) (1988) *Sustainable Environmental Management*, Belhaven, London.

United Nations (1991) *Demographic Yearbook 1989*, UN, New York.

United Nations (1991) *World Urbanisation Prospects 1990*, UN, New York.

United Nations Development Programme (1990) *Human Development Report 1990*, Oxford University Press, Oxford.

United Nations Development Programme (1991) *Human Development Report 1991*, Oxford University Press, New York.

United Nations Environment Programme (1989) *Environmental Data Report 1989/90*, Blackwell, Oxford.

Vitousek, P., Ehrlich, P. R., Ehrlich, A. H. and Matson, P. (1986) 'Human appropriation of the products of photosynthesis', *Bioscience*, 36(6), pp. 368–373.

Wann, D. (1990) *Biologic*, Johnson Books, Boulder.

Webb, A. (1991) *The Future for U.K. Environment Policy*, The Economist Intelligence Unit, Special Report 2182, London.

Ward, B. (1979) *Progress for a Small Planet*, Penguin, Harmondsworth.

Webster, F. E. (1974) *Social Aspects of Marketing*, Prentice-Hall, Englewood Cliffs.

Weiner, J. (1990) *The Next Hundred Years*, Bantam, New York.

Weiss, E. B. (ed.) (1992) *Global Environmental Change*, United Nations University Press, Tokyo.

Welford, R. and Gouldson, A. (1993) *Environmental Management and Business Strategy*, Pitman, London.

Wells, P. and Jetter, M. (1991) *The Global Consumer*, Victor Gollancz, London.

Westoby, J. (1989) *An Introduction to World Forestry*, Blackwell, Oxford.

Weston, J. (ed.) (1986) *Red and Green*, Pluto Press, London.

Wheale, P. and McNally, R. M. (eds) (1990) *The Bio-Revolution*, Pluto Press, London.

Wheelwright, S. C. and Clark, K.B. (1992) *Revolutionising Product Development*, Free Press, New York.

Whitehead, A. N. (1925) *Science and the Modern World*, Macmillan, New York.

Whiteley, R. C. (1991) *The Customer Driven Company*, Addison-Wesley, Reading.

Wilson, E. O. (ed.) (1988) *Biodiversity*, National Academy of Sciences, Washington DC.

Wilson, E. O. (1992) *The Diversity of Life*, Harvard University Press, Cambridge.

Winpenny, J. (1991) *Values for the Environment*, HMSO, London.

Winter, G. (1988) *Business and the Environment*, McGraw-Hill, Hamburg.

Womak, J. P., Jones, D. T. and Roos, D. (1990) *The Machine that Changed the World*, Rawson Associates, New York.

Wood, B. (1984) *E F Schumacher*, Harper and Row, New York.

World Bank (1988) *Social Indicators of Development 1988*, Johns Hopkins University Press, Baltimore.

World Bank (1990) *Poverty*, Oxford University Press, Oxford.

World Bank (1991) *The World Bank and the Environment*, The World Bank, Washington DC.

World Resources Institute (1992) *World Resources 1992 – 1993*, Oxford University Press, New York.

Worster, D. (1977) *Nature's Economy*, Cambridge Unviversity Press, Cambridge.

Worster, D. (1987) *Nature's Ecology*, Cambridge University Press, Cambridge.

Wrigley, N. (ed.) (1993) *Store Choice, Store Location and Market Analysis*, Routledge and Kegan Paul, London.

Wurman, R. S. (1991) *Information Anxiety*, Pan, London.

Yankelovich, D. (1981) *New Rules*, Randon House, New York.

Young, J. (1990) *Post Environmentalism*, Belhaven, London.

Young, J. E. (1991) *Discarding the Throwaway Society*, Worldwatch Paper 101, Worldwatch Institute, Washington DC.

Zenisek, T. J. (1979) 'Corporate social responsibility', *Academy of Management Review*, 4(2), pp. 359–369.

Zimmermann, E. W. (1933) *World Resources and Industries*, Harper and Brothers, New York.

Zuboff, S. (1988) *In the Age of the Smart Machine*, Basic Books, New York.

Index